AUSTRALIA'S TRADE POLICIES

Edited by

Richard Pomfret

Melbourne
OXFORD UNIVERSITY PRESS
Oxford Auckland New York

OXFORD UNIVERSITY PRESS AUSTRALIA
Oxford New York Toronto
Delhi Bombay Calcutta Madras Karachi
Kuala Lumpur Singapore Hong Kong Tokyo
Nairobi Dar es Salaam Cape Town
Melbourne Auckland Madrid
and associated companies in
Berlin Ibadan

OXFORD is a trade mark of Oxford University Press

First published 1995

National library of Australia
Cataloguing in publication data:

Australia's Trade Policies.
Bibliography.
Includes index.
ISBN 0 19 553536 7
1. Australia — Commerce. 2. Australia — Commercial policy.
1. Pomfret, Richard W.T.
380.10994

Edited by Cathryn Game
Indexed by Jodie Nelson
Cover design by Modern Art Production Art
Typeset by Dobell, Temple & Associates
Printed by Australian Print Group
Published by Oxford University Press
253 Normanby Road, South Melbourne, Australia

Contents

Figures

Tables

Acknowledgments

This book emerged from a conversation in 1992 in which Alan Fettling of Oxford University Press mentioned to Tom Sheridan of Adelaide University's Economics Department that there was a need for a book on Australian trade policies. Tom suggested that the Centre for International Economic Studies could produce such a book. I, as acting director of the CIES, during Kym Anderson's secondment to GATT, agreed to edit the volume.

I am grateful to Alan Fettling for his support and encouragement, and unsure whether I should be grateful to Tom Sheridan. The book has been a collaborative project by CIES associates; my thanks go to all the authors for their cooperation, and especially to Kym Anderson for undertaking editorial duties when I was absent and to Ian McLean for taking on extra proofreading. At a workshop on draft chapters in July 1993 John Siegfried gave many useful suggestions. Two anonymous readers provided helpful comments on the penultimate draft.

Richard Pomfret

Contributors

Kym Anderson, professor of economics and director of the Centre for International Economic Studies at the University of Adelaide, has research interests in trade and other distortionary policies as well as in agricultural, development and environmental economics.

Bijit Bora, lecturer in economics at the Flinders University of South Australia, is an international economist with particular interests in the economics of foreign investment.

Christopher Findlay, associate professor of economics and co-director of the Chinese Economy Research Unit at the University of Adelaide, focuses his research on trade in services and in minerals and energy, and on the Chinese economy.

Ian McLean, reader in economics at the University of Adelaide, specialises in examining Australia's economic history in comparative and international perspectives.

Jonathan Pincus, George Gollin professor of economics and head of the Department of Economics at the University of Adelaide, has a wide range of interests including economic history, federalism and constitutional economics.

Richard Pomfret, professor of economics at the University of Adelaide, writes on numerous aspects of trade and development economics including the transition of former centrally planned economies.

Colin Rogers, associate professor of economics at the University of Adelaide, specialises in open economy macroeconomics and monetary theory.

Abbreviations and Acronyms

APEC	Asia–Pacific Economic Cooperation
ASEAN	Association of South East Asian Nations
CAP	Common Agricultural Policy (of the EC/EU)
CER	Closer Economic Relations (agreement between Australia and New Zealand)
CTE	consumer tax equivalent
DFI	direct foreign investment
EAEC	East Asian Economic Caucus
EC	European Community
EU	European Union (formerly the EC)
GATS	General Agreement on Trade in Services
GATT	General Agreement on Tariffs and Trade
GDP	gross domestic product
GNP	gross national product
GSP	generalised system of preferences (for developing countries)
HECS	Higher Education Contribution Scheme
IAC	Industries Assistance Commission (successor to the Tariff Board) now called Industry Commission
ITO	International Trade Organisation (stillborn)
MFA	Multi-fibre Arrangement
MFN	most favoured nation
MNC	multinational corporation
NAFTA	North American Free Trade Agreement
NSE	net subsidy equivalent

OECD	Organisation for Economic Cooperation and Development
OPEC	Organisation of Petroleum Exporting Countries
PECC	Pacific Economic Cooperation Council
RTA	regional trading arrangement
SITC	standard international trade classification
TCF	textiles, clothing and footwear
UNTAC	United Nations Transitional Authority in Cambodia
VER	voluntary export restraint agreement
WTO	World Trade Organisation (result of the Uruguay Round)

Preface

Richard Pomfret

Trade policies have seldom been out of the headlines in Australia, but in the last quarter of the twentieth century the debate has been particularly open. At the end of the 1960s Australia's tariffs on imports were higher than those of every other high-income country except New Zealand. Beginning in 1973 and accelerating in the 1980s, Australia undertook a reduction in trade barriers unparalleled in her history and intended to make them among the lowest in the world by the end of the century. Since 1983 this dramatic policy reform has enjoyed bipartisan support from the major political parties, but is still treated with suspicion by much of the population. Concern about the unemployment consequences of the closure of Nissan's car manufacturing operations, combined with the decisive loss of a previously safe Labor seat in a 1992 by-election in which the main issue was the decline of the textile and clothing industry in the face of foreign competition, emphasised that twenty years after trade liberalisation began it still faced opposition.

Apart from the policy debate, dramatic changes in trade relations have influenced domestic society and the country's role in the world. The most obvious trade-led change has been the reorientation of Australia from the British connection, which was still dominant in the

1950s but had become economically minor by the late twentieth century. Displacing British influence has been an initial increase in links with the USA and then a huge redirection of trade towards Japan and other East Asian economies.

Trade policies and trade relations must be seen in a broader perspective. For much of Australia's history over the last two centuries the prime goal was to increase population, and this shaped many policies, including trade policy. In recent decades concern has shifted towards worrying about Australia's relatively slow increase in income per head. At the same time the shift in trade partners made the White Australia approach to immigration increasingly anachronistic and untenable. Thus, trade policies must be seen as just some tiles of a mosaic.

It is difficult to define trade policies precisely. To continue with the mosaic metaphor, it is unclear which tiles belong in the trade policy group and which do not. Taxes on imports or exports are obviously trade policies, as are quantitative restrictions on imports or exports, but other policies affect trade more or less directly. Health standards might, intentionally or not, rule out food imports from some countries, and detailed roadworthiness standards (which are not standardised internationally) make it difficult for individuals to import cars into Australia. Policies to increase the degree of competition in domestic markets are more effective (and might even become unnecessary) if restrictions on imports are removed, while competition policies that promote or hinder efficiency will affect the competitiveness of Australian producers as exporters. Labour market conditions that increase firms' costs reduce their ability to compete in export markets and so forth. Indeed, every policy that differs in implementation between countries or differs in its impact across economic activities can have some impact on trade flows. Later chapters will discuss the most important of these non-tariff barriers to trade but will also try to draw a line between policies more directly affecting international trade and policies with more indirect effects.

The aim of this book is to provide a self-contained analysis of Australian trade policies. The intention is to make the book accessible to as large a range of interested readers as possible, although the biggest market will be students in second- or third-year economics courses at Australian universities. These students will already have studied economic theory and have had some contact with international trade theory. To cater for readers' divergent backgrounds in economics the introduction offers a terse but complete review of all the theoretical knowledge required to understand subsequent chapters.

Chapters 1 and 2 set the scene and provide the empirical and historical basis for the book. Australia has always been an open economy, maintaining a high living standard on the basis of its abundant endowment of natural resources. Since the second half of the nineteenth century Australian governments have used trade policies to redistribute some of the rents from the natural resources and to diversify the economic structure. Chapter 1 describes the evolving structure of production and trade, and compares Australia's economic performance with that of other countries. Chapter 2 provides a chronological account of the evolution of Australian trade policies while also seeking to identify the political economy forces that have shaped this evolution.

The first two chapters identify the major questions to be addressed in this book and offer some initial answers. A fundamental question concerns the importance of international trade and trade policies as determinants of economic growth and well-being. Other recurring questions include the following. Why did Australia introduce trade barriers and keep them so high for so long? Once Australia abandoned protectionist tariffs, why were they reduced so quickly, despite two recessions in the 1980s? As tariffs and other artificial trade barriers are reduced, to what extent will Australian producers face greater competition from imports, and to what extent will they still be sheltered by the natural protection of distance? What is the relationship between trade policy and other elements of the economic liberalisation program, especially the freeing of capital markets? How far is Australia affected by other countries' trade policies, and can Australia do anything to influence those policies? What does the future hold for Australian trade policies and performance?

Chapters 3–5 analyse Australian trade policies in more detail by distinguishing primary, manufacturing and service sectors. Historically, Australia's international trade has been determined by her role as an exporter of primary products and importer of manufactured goods, and policies have been intended to soften this dependence on imports of manufactures. Important developments during the second half of the twentieth century have been a major shift in the composition of primary exports and fluctuations in Australia's terms of trade as the prices of primary products have risen and fallen. Meanwhile, since the early 1970s Australian governments have reversed their attempt to protect manufacturers from competing imports, although two branches of manufacturing have been partly exempted from this liberalisation. Chapter 4 analyses the effects of reduced protection and of the special protection granted to the automobile and textile, clothing and footwear branches.

Before the 1980s trade policies were considered to be almost exclusively concerned with trade in goods. An innovative element of the Uruguay Round of trade negotiations, initiated in 1986, was the inclusion of services. This has been important for Australia because the service sector accounts for a large part of economic activity, and some parts have been deregulated since the early 1980s. One part of deregulation has been liberalisation of services trade with New Zealand under the Closer Economic Relations agreement, but more generally Australia's trade in services has been growing rapidly. Trade in airline services, financial services and educational services have all been in the headlines in recent years. The pattern of Australia's trade in services and the policy issues raised are analysed in chapter 5.

Ever since European settlement began, Australia has been an open economy relying heavily on foreign labour and capital. Chapter 6 analyses the relationships between the flows of these factors of production and trade flows. Trade policies affect factor flows, and policies towards the inflow of labour and capital will also affect trade. During the 1970s and 1980s there were dramatic changes in both labour and capital flows as Australia abandoned the White Australia policy of immigration and as foreign exchange and capital markets were deregulated.

Deregulation of capital markets created new links between trade policies, which are essentially microeconomic (applying to individual activities), and macroeconomic variables. Chapter 7 analyses these links, focusing on the sequencing issue of whether Australia was well served by the policy of liberalising capital flows near the beginning of the trade liberalisation program. Chapter 7 reinforces the general argument that trade policies have little impact on macroeconomic variables like total output or aggregate unemployment in the medium term, but macroeconomic policies can influence the political support for trade liberalisation adversely if rising unemployment is seen to coincide with the reduction of trade barriers.

Australia is unique in that all of her major trading partners are far away. The closest international neighbour to Australia's main population centres is New Zealand but, because the two countries have similar resource endowments, trade between them has been fairly minor for both sides. Nevertheless, relations across the Tasman have some interesting policy aspects. New Zealand in the 1880s seriously considered whether to join the Austral(as)ian federation, but decided not to participate in the new customs union. How beneficial was the decision to remain outside the Australian tariff wall, pursuing independent trade policies, and would other colonies (e.g., Western Australia) have done better to have followed this lead? During the 1960s Australia and New Zealand signed a free trade agreement, which could have reversed some of the consequences of New Zealand's independent trade policy but in practice had little impact. Then in 1983 trans-Tasman relations were stimulated by the Closer Economic Relations agreement, which many observers see as a path-breaking measure going beyond narrowly defined trade policies and pointing the way that global trade agreements will follow in the twenty-first century. Chapter 8 addresses all of these issues.

The question of the international environment in which Australia trades is taken up in chapter 9. Although this book focuses on Australian trade policies, Australia has always been strongly affected by other countries' policies. Chapter 2 shows how Australia suffered from the global shift towards protection in 1930 and benefited from the global trade liberalisation after the signing of the General Agreement on Tariffs and Trade (GATT) in 1947, even though both of these developments were outside her control. Although Australia is too small a trading nation to exert much influence on the policies of the largest countries, it has sufficient weight to play a role in international economic diplomacy; measured by gross domestic product, Australia is the world's seventh largest economic unit after the European Union, the USA, Japan, Canada, China and Brazil. During the 1970s Australia belatedly began to participate more actively in international trade negotiations under GATT and in the 1980s started to play a leading role in agricultural negotiations through the Cairns Group. Meanwhile concerns began to grow about signs that some of the major trading nations might be retreating from open multilateral trading relations and forming regional blocs in Europe and in North America, with negative consequences for countries like Australia that had no bloc.

Regionalism versus multilateralism is just one of the important new trade issues that have arisen in recent years. Trade in services and investment-related issues are others. Some, like growing concern about the environment, are just peeping over the horizon. The epilogue aims to evaluate the future of Australian trade policies, placing it in a global context of evolving concerns. The implications will surely spill over into foreign policy and Australia's view of its position in the world. Will Australia continue to play a greater leadership role, both internationally and in the Asia–Pacific region, or were the high profile of the Cairns Group in the 1980s and of the Japanese–Australian-led successful supervision of the 1993 Cambodian election temporary aberrations from a long-term commitment to hedonistic minding of one's own business?

7-27

Introduction: The Economics of Trade Policy

F13

Richard Pomfret

International trade theory is one of the oldest and best-developed branches of economic theory, so it is natural to expect that theory will provide many guidelines to understanding the impact of trade policy in Australia. This chapter will present an introduction to trade theory, with emphasis on the tools and results most often used in the remainder of the book.[1]

The gains from trade

The most fundamental concept in international trade theory, that of comparative advantage, was first developed by David Ricardo almost two centuries ago. Although it is obvious why two countries might trade commodities that one can produce and the other cannot (e.g., Saudi Arabia trades oil for grain), it is less self-evident why countries might trade commodities that they can both produce. Ricardo's explanation was that as long as opportunity costs differ there are gains from specialisation and trade.

Even if one country is the most efficient at producing all goods, there are benefits from specialising in the goods in which its relative efficiency is greatest, i.e., those in which its comparative advantage lies. Since any two countries' opportunity costs are likely to differ for many pairs of goods, comparative advantage offers a pervasive explanation for trade, which is supported by the fact that all countries engage in international trade. Moreover, since specialisation yields greater total output, the potential exists for trade to be mutually beneficial.

Comparative advantage is a powerful explanation of trade, because situations in which trade is the only way to obtain a good are rare. Most countries could produce practically any commodity if they put enough resources into that activity. Russia could send people into space and stock a nuclear arsenal, but at the huge cost of diverting resources away from the production of consumption goods that would have improved people's daily lives. Even primary products can often be produced under adverse conditions, e.g., the Netherlands grows tomatoes by using energy to heat greenhouses while tomatoes can be produced at much lower resource cost in Italy where sunshine is plentiful and free. The relevance of this point for Australia is that Australia could produce a whole range of manufactured goods, but pursuing comparative advantage and trading primary products for manufactured goods that would be costly to produce domestically might provide Australians with more of everything.

Ricardo (1817, chapter 7) illustrated the gains from trade by using the example of a hatmaker and a shoemaker:

> Two men can both make shoes and hats, and one is superior to the other in both employments; but in making hats he can only exceed his competitor by one-fifth or twenty per cent, and in making shoes he can excel him by one-third or 33 per cent; will it not be for the interest of both, that the superior man should employ himself exclusively in making shoes and the inferior man in making hats?

More generally, if the slopes of the product transformation curves (also known as production possibility frontiers) for any pair of commodities differ between two countries, then there are gains from trade. In figure 0.1, country A is more efficient at producing wheat and country B is more efficient at producing cloth. By specialising and trading the combined production of wheat and cloth is increased, and there are gains from trade as long as the increased output outweighs any added transport costs.[2]

Figure 0.1 Potential gains from trade with linear product transformation curves

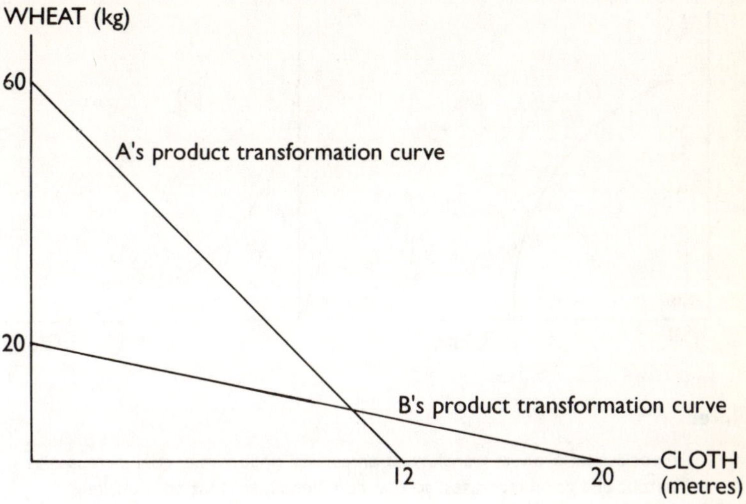

Note: The opportunity cost of producing a metre of cloth is 5 kg of wheat in country A and 1 kg of wheat in country B. In autarky each country will produce a mixture of output depending on domestic preferences, e.g. 8 m of cloth and 20 kg of wheat in A and 10 m and 10 kg in B. With specialisation their joint production could be 20 m and 60 kg, which with trade could permit higher consumption of both goods in both countries.

The linear product transformation curves in figure 0.1 represent constant opportunity costs. The more usual assumption is that opportunity costs increase as output of a good expands. This might be because of decreasing returns to scale or it might be because higher output pushes up the price of inputs used intensively in that activity. With the product transformation curves described in figure 0.2a, each country's comparative advantage depends not only on where it produces on its curve but also on where other countries are on their transformation curves. Countries will gain from trade as long as opportunity costs differ; i.e., gains from trade will exist until the slopes of their product transformation curves at the production point are identical.

In autarky, the production point and consumption point must

Figure 0.2 Non-linear product transformation curves

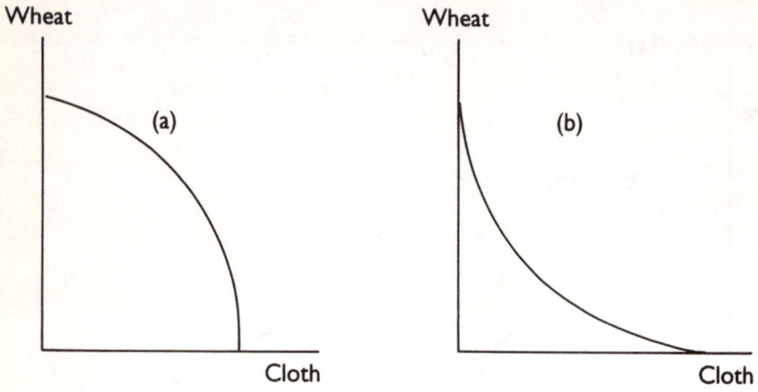

Notes

(a) Neoclassical product transformation curves: opportunity cost increases as output of a good increases, so that complete specialisation is unlikely.
(b) Strong economies of scale: trade will most likely lead to a corner solution.

coincide. One way to introduce demand-side forces is to represent a nation's preferences by a social welfare function. The social indifference curves associated with this function trace out alternative combinations of goods that yield the same level of utility. Welfare is maximised by reaching the highest possible social indifference curve; given the economy's productive capacity, this is curve I_0 in figure 0.3. Competitive markets will achieve this outcome with a price ratio equal to the slope of both the indifference curve (i.e., the marginal rate of substitution) and the product transformation curve (i.e., the marginal rate of technical substitution, which is equal to the opportunity cost ratio).

If a market economy has a comparative advantage in a good, it is revealed by an autarky price ratio that differs from that in the world market. In figure 0.3 the world price line *WW* is flatter than the domestic price line *DD*. The country has a comparative advantage in wheat. By changing its output mix from Q_0 to Q_1 and trading at the world price, the country can consume at point C_1, which is on the higher social indifference curve I_1. Note that the benefit comes not only from specialisation in production but also from the separation of production

Figure 0.3 The gains from trade for a small economy

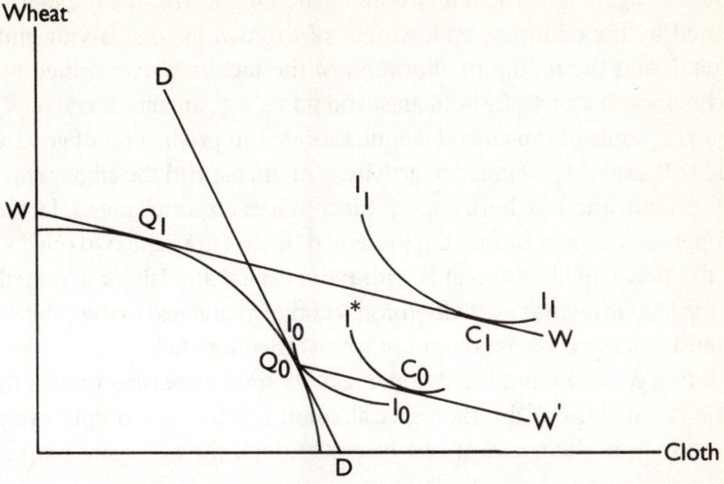

Note: Without trade, production and consumption are at point Q_0. With trade, the country specialises in wheat, producing at point Q_1, and can consume at any point along the world price line *WW*. C_1 yields the highest feasible social welfare. Even if production remains unchanged at Q_0, welfare can be increased by trading at world prices (selling wheat and buying cloth) to reach consumption point C_0, which is on a higher social indifference curve than Q_0.

and consumption; even if output were unchanged at Q_0 trade would allow the society to consume at C_0 on indifference curve I^*, which represents higher welfare than the indifference curve I_0.

The welfare effects of more trade might be negative if the country's actions can change world prices. In figure 0.3, as the country expands its wheat exports, the world price of wheat will fall (i.e., *WW* becomes steeper), reducing the benefits from trade. If the country is growing and the growth is biased towards the export sector, then downward pressure on export prices (i.e., deteriorating terms of trade) could offset the benefits of higher physical output, leading to immiserising growth (i.e., lower GNP valued at world prices).

If the country is too small to affect world prices, then growth due to increased factor inputs will have a magnified effect on the output mix. If growth is due solely to increases in one factor, then the output of the good using the non-growing factor relatively intensively will decline absolutely (and not just as a share of the bigger total output, as one might expect). This result, known as the Rybczynski Theorem, is illustrated in figure 0.4. The dimensions of the Edgeworth Box are determined by the country's endowment of the two factors, labour and capital, and the optimum allocation of the factors is determined by technology (portrayed by isoquants) and prices, e.g., at point A with O_xK_x and O_xL_x units of capital and labour allocated to production of good *x* and O_yK_y and O_yL_y allocated to activity *y*. An increase in the labour supply extends the box horizontally. Since prices are unchanged, factor proportions remain constant, represented by rays from the two origins, so the new equilibrium is at B with more capital and labour allocated to the labour-intensive activity (production of *x*) and less to the capital-intensity activity, whose output of *y* must therefore fall.

If two goods are produced under conditions of increasing returns to scale (as in figure 0.2b), then specialisation will increase output even more than in the constant cost case. Although the aggregate output

Figure 0.4 Effect on the output mix of growth in one factor

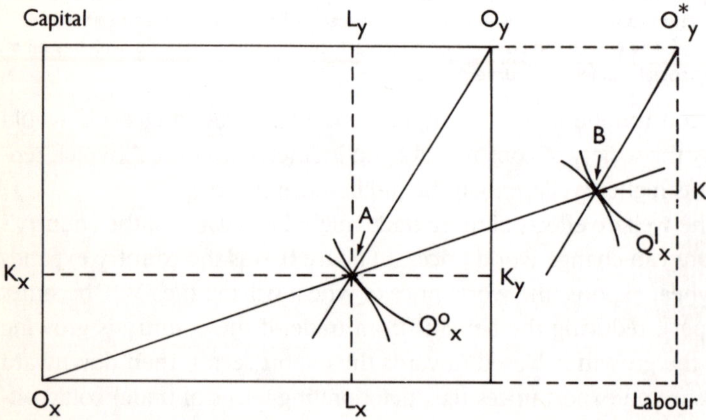

Note: The Rybczynski Theorem. After an increase in the labour force, output of the labour-intensive good *X* increases, that is, isoquant Q^I_x represents a higher output level than Q^0_x. Since the amount of capital allocated to *Y* production has fallen, from O_yK_y to $O^*_yK^*_y$, and the capital:labour ratio is unchanged, output of the capital-intensive good *Y* must fall.

gain is larger in the presence of scale economics, the world price might end up being such that one good's price is lower than it was in either country before trade. Thus trade might not be mutually beneficial; the country producing the good whose price falls could, due to the terms of trade effects, be worse off than if it did not trade. There is no evidence of this theoretical possibility having occurred and, for Australia, these arguments against trade are of limited relevance.

In sum, there are overwhelming benefits to Australia from international trade. This is why, since the arrival of the Europeans, Australia has always participated in international trade and been an open economy. Trade policy debates have never been over whether to trade, but rather over whether to intervene at the margin to encourage or discourage trade in certain products. The microeconomic nature of trade policy measures has also meant that debates have often focused on the distributional consequences of more or less trade, which are hidden by using a social welfare function but to which we will return in the third section of this introduction.

The pattern of trade

The existence of gains from trade depends on differences in opportunity costs or in tastes. Explaining the pattern of trade means identifying the reasons for such differences. Economists have had little to say about tastes, and generally assume them to be similar across countries or at least not to be a significant determinant of trade patterns. Australia exports Vegemite for overseas addicts facing withdrawal symptoms, but it is not a major trade item.

The simplest, but least enlightening, explanation of differences in opportunity costs is to assume different capabilities. Italians have a comparative advantage in fashion clothes because they have a highly developed aesthetic sense. New Zealanders export rugby shirts because they are good at rugby.

A more insightful explanation of trade patterns focuses on differences in factor endowments. If wheat uses land relatively intensively and cloth uses labour relatively intensively, then a land-abundant country will have a flatter product transformation curve than a labour-rich country (see figure 0.2a). For given tastes and technology, each country's comparative advantage lies in the good that uses most intensively the factor of production in which the country is relatively well endowed. This result is known as the Heckscher–Ohlin theorem.

The Heckscher–Ohlin theorem can be extended beyond two factors, although it then becomes difficult to rank goods by their factor intensity. By casual observation, it is apparent that India exports labour-intensive goods, Japan exports capital-intensive goods and Australia exports land-intensive goods. Although no simple theorem can capture all trade patterns, the factor proportion explanation has considerable empirical support (e.g., Leamer 1984). Apart from natural resources, Australia's comparative advantage lies in goods requiring specific skills, and her comparative disadvantage is clearly in goods whose main input is unskilled labour (see figure 1.1).

In some cases it is difficult to explain trade patterns by factor endowments, especially when there is intra-industry trade. If the same good is simultaneously imported and exported, then such trade is incompatible with the Heckscher–Ohlin theorem (a good cannot be intensive in both the country's abundant factor and its scarce factor). Some studies have found high levels of intra-industry trade, but there is considerable disagreement over the significance of these findings, and estimated intra-industry trade is lower for Australia than for most other countries.[3] Intra-industry trade can be a result of product differentiation, e.g., Australia exports but also imports wines; although the goods fall under the same commodity heading, presumably drinkers consider a Medoc and a Coonawara to differ. The key issue is how to define a good, and the implication is that detailed case studies (as in the later chapters of this book) are necessary to understand trade patterns in more complex product categories.

TRADE AND INCOME DISTRIBUTION

The gains from trade described above will not be distributed equally among the population. In principle, if there are net benefits, then the losers could be compensated by the gainers, but such redistribution is likely to be opposed by the gainers and to be difficult to implement in any case. This is important because the losers from any policy change will oppose it, and if they are sufficiently powerful they might prevent the pursuit of policies which are in the national interest.

The two-good general equilibrium model in figure 0.3 has already pointed to one distributional effect of trade. Production of export goods expands with trade while production of import-competing goods contracts. In the short run workers and owners of capital or land used in the import-competing activities face adjustment costs, and some skills or equipment might become worthless. As the relative price of imported goods falls, consumers of those goods benefit, while consumers of export goods suffer from higher relative prices.

Restricting trade has the opposite effects to those described in the previous paragraph. The short-term impacts on producers and consumers of the import-competing good can be illustrated by a partial equilibrium analysis. Figure 0.5 shows the effects of imposing a trade restriction that introduces a wedge between the domestic price and the world price of an imported good. With free trade, OJ units are produced and MJ units are imported. A tariff increases the domestic price by the size of the tariff (P_d–P_w), reducing consumption to OL, increasing domestic output to OK and cutting imports to KL. By specifying a 'small' country, it is assumed that the change in imports has no impact on the world price. Consumer surplus is reduced by area a + b + c + d, while producer surplus increases by area a. Area c accrues to the government as tariff revenue. Thus, loss to domestic consumers is partially offset by gains to producers and revenue to the government. Giving equal weight to all welfare changes, there is a net welfare cost measured by the two triangles *b* (the production-side loss from producing domestically units which could be imported at lower resource cost) and *d* (the consumption-side loss from preventing consumers from purchasing units LM at their true opportunity costs).[4] The net costs are simply the forgone gains from trade, but figure 0.5 suggests that the internal redistribution effects of trade policies are likely to be far larger than the net gains or losses.

Expansion of trade will increase the demand for inputs used intensively in the export industry. In the neoclassical trade model with competitive markets and fixed factor endowments, the scarce factor will lose and the abundant factor gain from bringing the domestic price closer to the world price. Conversely, sheltering the import-competing sector behind trade barriers will benefit the scarce factor and hurt the abundant factor. This result, the Stolper–Samuelson theorem, is reflected in the protectionist tendencies of trade unions in high-income countries and of capitalists in low-income countries.

Figure 0.5 Partial equilibrium analysis of a tariff imposed by a small country

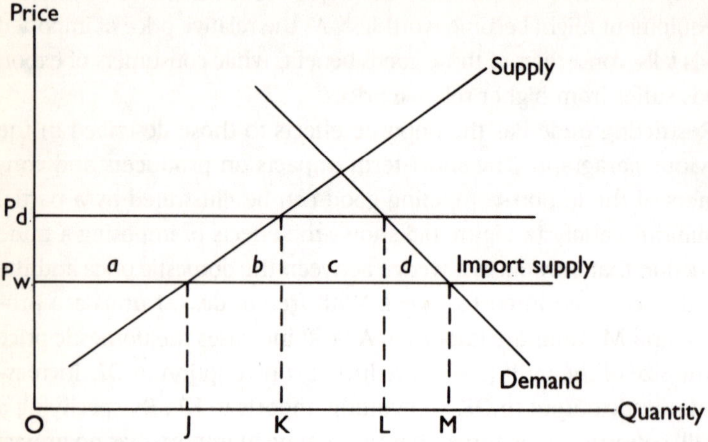

Note: A tariff drives a wedge between the domestic price (P_d) and the world price (P_w).Consumer surplus is reduced by area $a+b+c+d$, while producer surplus increases by area a and area c accrues to the government as tariff revenue. The net welfare cost to the importing country is measured by the two triangles b and d.

Thus, there are two dimensions to the income distribution effects of trade restrictions. Restricting imports will benefit producers of import-competing goods and hurt consumers of those goods. The trade barrier will also benefit the scarce factor of production at the expense of the abundant factor. Which effect dominates depends on the time horizon and degree of factor mobility between sectors. In the short run the partial equilibrium effect will dominate, but in the long run the Stolper–Samuelson result will come into play. The more factors are specific to their current use, the bigger the historical time difference between the short run and the long run. If owners have capital specific to their import-competing industry, then they will oppose removal of import restrictions even if they recognise that trade liberalisation will ultimately benefit capital as the abundant factor in Australia.

If the world were characterised by perfectly competitive markets, then even in the absence of international factor mobility the free movement of goods would equalise factor prices globally. Factor price equalisation has obviously not happened, although there is evidence that trade does contribute to some convergence. In practice, factors are mobile internationally, although there remain more serious restrictions than exist in goods markets, and factor mobility provides an additional mechanism promoting goods and factor price equalisation in an integrated world economy.

Reasons for intervention

Most common arguments in favour of interventionist trade policies are based on some kind of externality, which means that social costs and benefits are not fully reflected in market prices. The externality might lead to domestic output being below the socially desirable level, or it might involve above optimal consumption or below optimal government revenue. In all of these cases a trade barrier could bring the economy closer to the optimum, although there will usually be a superior way of achieving this goal.

Manufacturing industries have frequently been argued to possess characteristics favourable to long-term economic development. Many governments have tried to promote particular industrial activities considered to have wide-ranging linkage effects, e.g., steel production or, more recently, computer-related activities. Some industries are associated with economic maturity, and governments are eager to ensure that they are nurtured, e.g., automobiles. Australian governments have historically been rather unselective, supporting manufacturing in general, although they have been particularly concerned about the car industry. Trade barriers can clearly contribute to achieving such goals because they increase domestic output of protected industries (as in figure 0.5).

Similar arguments can be made about consumption externalities. Some goods are perceived to have negative externalities, and trade barriers will help to reduce consumption. Luxury goods, alcoholic beverages, narcotics and pornographic material are the most common items in this category. Import restrictions, often total bans, are a way of increasing the domestic price and reducing consumption below the free market level.

A related argument could be used to justify tariffs as a source of government revenue. All revenue sources have associated costs, either in costs of collection or resource misallocation costs resulting from distorted incentives. Tariffs are one of the easiest taxes to collect, especially for an island economy, so that the negative side-effects might be lower than for other sources of government revenue.

Some of these arguments for protection from imports might emphasise the importance of time. The infant industry argument is the best known. Since it takes time to establish production and sales of some goods, potential producers might be discouraged by the fact that they will initially be unable to compete with imports, even if the country has a comparative advantage in the good which would ensure that the project would eventually be profitable. As long as the discounted stream of net social benefits is positive, then there is a case for providing support for the industry in the short run, if the industry would be stillborn without support and if the support is removed when the industry grows up.

There is also a senescent industry argument for protecting activities that have lost an earlier comparative advantage. If the sudden decline of an import-competing industry would involve serious social costs, e.g., because it is concentrated in a region with few alternative employment opportunities, then the decline in output and employment can be delayed by imposing trade barriers. More generally, all changes in comparative advantage will involve changes in the output mix and the adjustment costs can perhaps be alleviated by temporary increases in trade barriers.

All of these arguments can be related to figure 0.5. Restricting imports of a good will encourage domestic production and reduce domestic consumption, as well as benefiting domestic producers and hurting domestic consumers; if restriction is by a tax on imports, then it will also raise revenue for the government. Introducing externalities is simply saying that a dollar's worth of consumer surplus or of producer surplus or of government revenue, when measured at market prices, should not be given equal weight. The subtraction exercise that leads to $b + d$ being the net welfare loss is no longer appropriate, and any of the above arguments for protection could be valid in a particular case.

One problem, however, is that even if there is no externality, beneficiaries of the trade restriction have an interest in emphasising the social value of the policy. Therefore, domestic producers faced by competition from imports tend to support import restrictions but to express their position in terms of the 'national interest' rather than in terms of the bigger profits they will make. Sometimes it will be difficult for policymakers to distinguish between private and public interest if the producers control relevant information about the nature of the industry.

The second and more conclusive problem is that, even if the externalities listed above do exist, trade policies are never the best way of dealing with them. The problem is not that trade policies are ineffective but rather that they do too much. If, for example, it is desirable to increase domestic output from *OJ* to *OK* in figure 0.5, then there is no way of avoiding the production-side loss of forgone gains from trade at market prices (triangle *b*), but it is not necessary also to incur the consumption-side loss (triangle *d*). The superior policy is to tackle the externality directly, in this case by subsidising domestic output; a subsidy of P_d–P_w per unit would have the same effect on production as the tariff but without distorting consumption. Similarly, if consumption is too high, then imposing a sales tax will reduce consumption without sacrificing the gains from specialisation in production. If adjustment costs are too high, then provide adjustment assistance, retraining programs and so forth. If an infant industry will be profitable in the long term, then provide a subsidy or, better still, reform capital markets so that an entrepreneur can borrow against future profits. In all of these cases, the superior policy not only avoids undesired side-effects but it is also likely to encourage more rigorous consideration of the case for policy intervention. Trade restrictions pass on to consumers most of the costs of supporting industries with positive externalities, and individual consumers are usually too little affected or too poorly organised to protest if the burden is too large.

The general theory of distortions, therefore, concludes that any distortion should be tackled as close to its root as possible. Trade barriers are one means of overcoming domestic distortions, but they have limited justification; trade policies might be better than doing nothing, but they are never the best policy. The same assessment applies to the revenue argument for tariffs. When governments had few alternative revenue sources, tariffs might have made sense, but today with bureaucracies capable of collecting income, sales and other taxes the collection cost argument is less important, and the resource misallocation costs of tariffs make them a very inferior revenue source. Table 2.2 (p. 59) documents how the importance of trade taxes as revenue sources has fallen dramatically since the nineteenth century.

The only times when trade policy is the best policy is when the distortion is in international markets. Thus, if Australia has market power in a particular commodity, national welfare might be increased by restricting trade in that commodity in order to move world prices in Australia's favour. This applies to some primary products and is a justification, in principle, for organising marketing boards which will exploit Australian monopoly power most effectively, although the track record of such boards is poor. Again, it is important to judge the outcome in terms of the national interest, rather than benefits to sections of the population, and to consider the long-term implications, e.g., pushing up world prices of wool might encourage people to develop a taste for synthetic fibres. The Australian wool industry's attempts to influence world prices by limiting exports in the late 1980s and early 1990s suggest that even in wool Australia's monopoly power is not all that strong (see chapter 3).

In the economics literature of the 1980s considerable attention was given to trade in imperfectly competitive markets. Without perfect competition, the benevolent conclusions about the gains from trade might not hold. In particular, in oligopolistic markets trade policies could be used to change the strategic interaction among firms, shifting the monopoly rents in favour of domestic firms. There is no evidence for this theoretical possibility ever having occurred, and the industries around which the debate has centred (aircraft and semiconductors) are not significant parts of the Australian economy. Nevertheless, the debate has had echoes in Australia, and its relevance will be discussed in more detail in chapter 4.

Employment arguments have featured prominently in Australian trade policy debates. From this perspective, figure 0.5 can be misleading, because it is a partial equilibrium diagram. Protecting an industry will increase output, and presumably employment, in an industry, but it will have an impact on the labour market that is not captured in figure 0.5. Assuming competitive markets (and therefore full employment), figure 0.3 illustrates the general equilibrium effects; if the output mix is shifted in favour of the import-competing industry, then this must be at the cost of the export industry. The general point is that trade barriers are microeconomic measures; protecting an activity will preserve jobs in that industry, but only at the cost of making labour more expensive and hence reducing jobs elsewhere.

Protectionist arguments are usually strongest in times of high unemployment. Understandably, people are more concerned then with losing their jobs and fear the adjustment costs associated with finding new employment. Nevertheless, aggregate unemployment is a macroeconomic phenomenon, and should be dealt with by macroeconomic policy. This is a reinforcement of the distortions argument against trade barriers; in the short run trade barriers can help to control unemployment by reducing the frictional unemployment associated with changing the output mix, but in the long run they are clumsy tools for tackling overall unemployment. In general, trade policies can influence the distribution of jobs between activities (i.e. favouring protected activities over non-protected, export and non-traded activities) but not the number of jobs in the economy.

ALTERNATIVE INSTRUMENTS OF TRADE POLICY

Any policy that discriminates between domestically produced and imported goods will create a wedge between domestic and world prices, and usually have the consequences described in the previous section. The details of the consequences will, however, vary according to the instrument used. This section analyses the most common forms of trade barriers.

The simplest and commonest trade barrier is a tariff, or tax on imports. The impact of a tariff on a good is captured by figure 0.5, assuming that the good is of such minor importance that indirect effects via input markets or the balance of trade can be ignored. A tariff protecting a major industry would also have general equilibrium effects, drawing resources away from other activities (i.e., by making the domestic price line steeper than *WW* in figure 0.3 — but not as steep as *DD*, unless it were a prohibitive tariff).

Tariffs are either specific (so many dollars per unit imported) or *ad valorem* (a percentage of the value of the import) or a mixture of the two. At any moment the legal form is immaterial because any specific duty has an *ad valorem* equivalent and vice versa. There is a difference over time insofar as a specific duty becomes less burdensome if prices are rising. Specific duties encourage quality upgrading because a $5 duty on shirts would discourage imports of cheap shirts more than imports of expensive shirts, and they are also regressive, hurting poor people more than rich people.

Anti-dumping duties are a form of taxes on imports that have become pervasive in recent decades and are especially prevalent in Australia. They differ from normal tariffs in that the level of the anti-dumping duty is determined by reference to a foreign price level. Anti-dumping proceedings are initiated when a foreign supplier is selling in the Australian market at less than the fair price. The justification for levying a duty is that the foreign supplier might be engaging in predatory pricing aimed at driving domestic producers out of the market, after which the foreign supplier will exploit its market power by charging monopoly prices. This case is, however, extremely rare, and anti-dumping duties usually operate as trade barriers to protect domestic producers from competitive foreign suppliers. By involving a judicial process of reviewing foreign firms' costs in order to determine the level of the duty, anti-dumping duties incur additional welfare losses over the usual costs of a tariff, especially if they can be levied retroactively.

The nominal tariff is not the best guide to resource allocation effects when traded inputs are involved in a good's production. The amount of protection given to a domestic activity depends on how much higher the value-added can be for domestic producers than for their foreign competitors; i.e., the extent to which they can receive higher factor incomes or be less efficient in factor use. The protection on value-added is known as the effective rate of protection (*ERP*), and is measured by the difference between value-added at world prices (*VA*) and value-added at domestic prices (VA^*). The size of the ERP depends on the spread between the tariff on the output (t_0) and the tariff on inputs (t_i) weighted by their share in the product's value (a_i = the amount of the *i*th input used in $1 worth of output). Thus:

$$ERP = (VA^* - VA)/VA = (t_0 - \sum a_i . t_i)/(1 - \sum a_i)$$

If the value-added is small relative to total value, even a low nominal tariff can be associated with a high *ERP*; e.g., low instant coffee tariffs plus free trade in coffee beans make for high *ERP* for Australian instant coffee producers (which helps to explain why Brazil and Colombia do not export instant coffee).

Tariffs might discriminate among different sources of imports. Developing country exporters, for example, face a lower Australian tariff on some items than do European, Japanese or North American exporters. The welfare effects on the importing country of a discriminatory tariff reduction are ambiguous. On the one hand, the reduced tariff might allow imports from the favoured trading partner to displace domestic production or increase domestic consumption with beneficial welfare effects; this *trade creation* can be pictured as a downward shift in the P_d line in figure 0.5. On the other hand, the discriminatory tariff reduction might divert trade from a lower-cost supplier to the favoured trading partner; for domestic consumers the favoured import is cheaper because it is subject to a lower tariff, but for the importing country the least-cost supplier is still the least-cost supplier and the consumers' gain is less than the loss of government revenue — the difference is the *trade diversion* cost of the tariff. The net economic effect of a discriminatory tariff cut depends on the balance between the trade creation benefits and the trade diversion costs. The preferred trading partner unambiguously benefits, and the importing country might be willing to accept an economic loss in order to please or to gain influence over the favoured trading partner. From a global perspective, the trade diversion associated with discriminatory trade barriers is undesirable, and the General Agreement on Tariffs and Trade, which lays out rules on international trade, has a strong presumption against discriminatory trade policies.

Trade barriers might restrict the quantity of trade directly rather than working on the price as a tariff does. Under certain assumptions a quantitative restriction, or import quota, will have identical effects to a tariff.[5] Thus, figure 0.5 could equally be representing the effects of a specific tariff P_d–P_w or of a quota *KL*. If the government administers the quota by selling import licences at the market-clearing price, P_d–P_w then the welfare effects are identical. If the import rights are distributed in any other way, then the area *c* will either accrue to importers as rent (i.e., they gain from being able to import at the world price and to sell at the higher domestic price) or be dissipated in rent-seeking activities (i.e., people will use resources in order to gain the valuable import rights); in the former case the only difference to the tariff analysis lies in the distribution of the gains, while the quantitative restriction is inferior in the latter case because the net cost consists of part of c as well as $b + d$.

The equivalence between tariffs and quotas can break down under certain conditions, of which the most important for a small country like Australia arises when domestic producers would possess market power in the absence of competing imports. If imports are subject to a tariff, then domestic producers can raise prices, but only up to the world price plus the tariff. If imports are subject to a quota, then a domestic monopolist can set both quantity and price in order to maximise profits. In figure 0.6, with a quota *AB*, the domestic firm faces an effective demand curve *D′EFD*; below the world price the firm supplies the whole market, above the world price it supplies domestic demand minus *AB* units. Profits are maximised by producing *OC* units (at which marginal cost equals marginal revenue) and charging price P_m. The net welfare cost from a quota is larger than that from an equivalent tariff because, in addition to the usual triangles $b + d$, there is the deadweight loss from monopoly, measured by the shaded triangle. Free trade or a non-prohibitive tariff is the best anti-monopoly policy, while an import quota allows monopolists to exercise their power over at least part of the market (and foreign suppliers will free-ride by charging price P_m for the *AB* units of imports).

Voluntary export restraints are a variant of quotas and are administered by the exporter at the request of the importing country. The exporting country typically agrees to a VER under heavy pressure to comply or face tougher trade barriers. The analysis is identical to that of an import quota except that the quota rent is more likely to accrue to the exporter, so that the net welfare effect on the importing country is even more negative. The popularity of VERs since the mid 1970s reflects the unwillingness of importing countries (especially the USA and EC) explicitly to break obligations not to increase tariffs or use import quotas, while being unable to resist protectionist pressures, especially in the textile and clothing, steel and automobile industries, where VERs have proliferated internationally.

Although quotas can be no better than (and are often much worse than) the equivalent tariff from a national perspective, producers often prefer a quota or a VER. The reasons are obvious when they have monopoly power or receive some of the import licences. Quotas are also preferred because, in situations where domestic costs are rising,

Figure 0.6 An import quota with a domestic monopoly

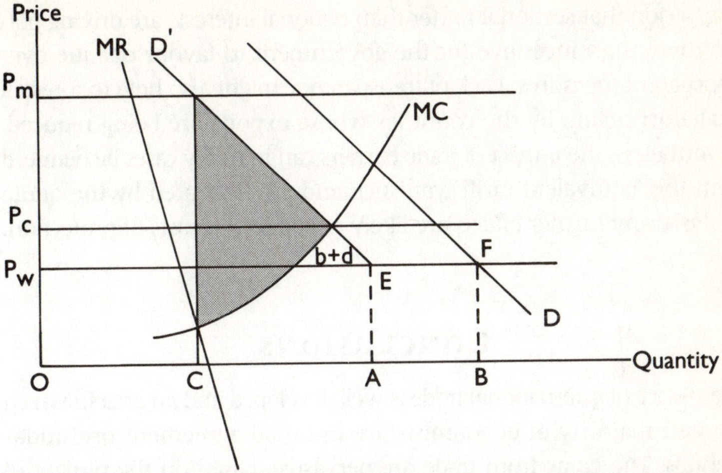

Note: With an import quota set at *AB* units, a domestic monopolist faces a demand curve *D'EFD*, and its profits are maximised at output *OC*, where marginal cost equals marginal revenue; at this output the firm charges a price P_m. With a tariff the domestic firm would be unable to charge above P_c, and the net social cost, $b + d$, would be the same as with a competitive industry, but with a quota there is the additional loss (measured by the shaded triangle) on the units which are no longer consumed but which purchasers value above their social cost of production.

world prices are falling or domestic demand is growing (all plausible in activities seeking protection), the equivalent tariff will be increasing over time; therefore a quota will provide more protection as it is 'needed', whereas a tariff will require continuous requests for increases, which might put the industry in a worse light and which might be refused.

Other non-tariff barriers are manifold but difficult to generalise about. Any technical, safety, health or other standard that is not harmonised internationally might favour domestic producers, who tend to design their goods for the Australian market, over foreign producers, for whom Australian sales are a small part of the total. Some standards are of course desirable, but there are cases of their being tailored to impose significant costs on foreign suppliers; it is often a fine line between having a minor incidental impact on trade and being a non-tariff barrier to trade. Local content laws and other performance requirements also act as non-tariff barriers, which need to be analysed in their specific context (see chapter 4 for an example).

Trade policy in the 1990s is seldom simple. Especially when there is a suspicion that sectional rather than national interests are driving policy, there is an incentive for the government to favour opaque over transparent measures. Lack of transparency might also help to forestall retaliatory action by the countries whose exports are being reduced. Nevertheless, the impact of trade barriers can in many cases be deduced from the 'equivalent tariff' yardstick and, as illustrated by the quota analysis, any further effects are likely to add to the costs of protection.

CONCLUSIONS

The theory of international trade is well developed and an area in which the vast majority of economists are in broad agreement on fundamentals. The gains from trade are pervasive, based on the pursuit of comparative advantage. Although little has been said here about the division of the gains from trade, it is crucial to note that specialisation and trade is a positive-sum game in which all participating nations can gain — and the overwhelming evidence is that they do all gain, with the precise distribution of the gains depending on elasticities of demand and supply.

The distribution of the gains from trade within a country is well understood but more controversial. Some people will lose from any change in trade policy unless they receive compensation from the gainers. This is the major reason why trade policy debates have such a high political profile, especially in a small open economy where most goods producers are either exporters or face competition from imports.[6]

In this brief survey I have not mentioned capital flows or macroeconomic aspects (trade is assumed to balance), which are usually excluded from pure trade theory and which are covered in chapters 6 and 7 of this book. Also omitted is analysis of the relationship between trade policy and economic growth. Since the gains from trade can be used to increase investment in physical or human capital as well as for consumption, there is a presumption that trade will increase the growth rate. During the 1950s and 1960s this impact was thought by many people (especially in countries exporting primary products) to be offset by the possibility that trade might lead to specialisation in sectors that had inherently less growth potential. More recently, the consensus has shifted towards emphasising less tangible but potentially important linkages between trade and growth such as exposure to new ideas and best-practice technology. The positive correlation between countries' openness and economic growth in recent decades provides empirical support for a positive link between trade and growth, but analysis of these relationships has yet to be integrated into mainstream trade theory, and the policy implications remain hotly debated.[7]

There are many valid reasons why interventionist trade policies might improve national welfare, but in almost all cases superior methods of achieving the desired objective exist. Therefore, in ranking policies to deal with a perceived problem, non-trade measures will usually be best, followed by a tariff, with other trade barriers last. In the simple and common case of choosing between a subsidy, a tariff and an import quota to stimulate an import-competing industry, for example, the social desirability of the three instruments is in that order.

In a poorly developed economy where implementation of alternative superior policies might not be feasible, then an interventionist trade policy might sometimes be better than doing nothing, but the theory of distortions provides a powerful case for a non-interventionist trade policy in late twentieth-century Australia. The theory of distortions effectively separates the argument for free trade from more general arguments for *laissez-faire* economic policies; one does not have to believe in non-interventionism and leaving everything to market forces in order to accept the benefits of free international trade.

1

AUSTRALIA'S CHANGING TRADE PATTERN AND GROWTH PERFORMANCE

Kym Anderson

This chapter paints an empirical picture of the changes in the structure and growth of the Australian economy and points to some of the ways trade policies have influenced those developments. It begins by examining how Australia's comparative advantages have evolved, especially since the 1960s, and how that, together with the changes in trade and industrial policies in Australia and elsewhere, has affected the composition and direction of the country's trade. In the light of its comparative advantages and policies at home and abroad, the chapter then assesses how well the Australian economy has performed in absolute terms and relative to the rest of the world. Attention is directed not only to output growth but also to changes in the industrial structure and location of production and employment. The chapter points to some trade-policy-related reasons for the relatively slow growth in *per capita* incomes in Australia, and then concludes by raising a number of pertinent questions that arise from this examination of the historical record, questions that are taken up in subsequent chapters.

The picture that emerges is an economy that is richly endowed by nature but one that has been making relatively poor use of its resources, not least because of its choice for many decades of a highly protectionist trade policy regime. One consequence of its policy choice is that

Australia's position as perhaps the world's highest-income country before World War I has slipped considerably, with it now struggling to remain in the top twenty list of high-income countries. Another consequence is that until recently its exports remained highly specialised in a small range of primary products.

The picture has begun to change though, as recent policy reforms — especially the lowering of import protection for non-primary producers — begin to have their expected effects. If the current reformist policy trends continue, the Australian economy's production and trade structure will be considerably different at the turn of the twenty-first century than a decade ago. Whether that will result in its *per capita* income growing much faster than it has historically, or only slightly faster, depends heavily on progress during the latter 1990s with other needed microeconomic reforms, most notably in the labour market.

THE DETERMINANTS OF AUSTRALIA'S COMPARATIVE ADVANTAGES

A country benefits from exporting goods that it can produce relatively cheaply and/or that are not strongly demanded at home, because with the proceeds it can import goods that are more cheaply produced abroad. To make use of such opportunities is to exploit one's comparative advantages. The sources of a country's comparative advantages can include differences between that country and the rest of the world in factor endowment ratios, technologies and consumer tastes. But, for natural-resource-rich, lightly populated Australia, the key determinant is the first of these, that is, its relative factor endowments. The reason this dominates all other determinants of Australia's comparative advantage is because the country is very richly endowed with agricultural land and mineral and energy resources per worker. Furthermore, for two centuries its workforce has been, by world standards, relatively highly skilled.

A useful way of summarising the relative resource endowments of different economies is provided in figure 1.1. That triangle, due to Leamer (1987), shows graphically various countries' relative endowments of three factors of production, denoted N for natural resources, L for labour time and C for produced capital (human, physical, knowledge,

Figure 1.1 Relative endowments of natural resources, labour and produced capital, various economies, 1991

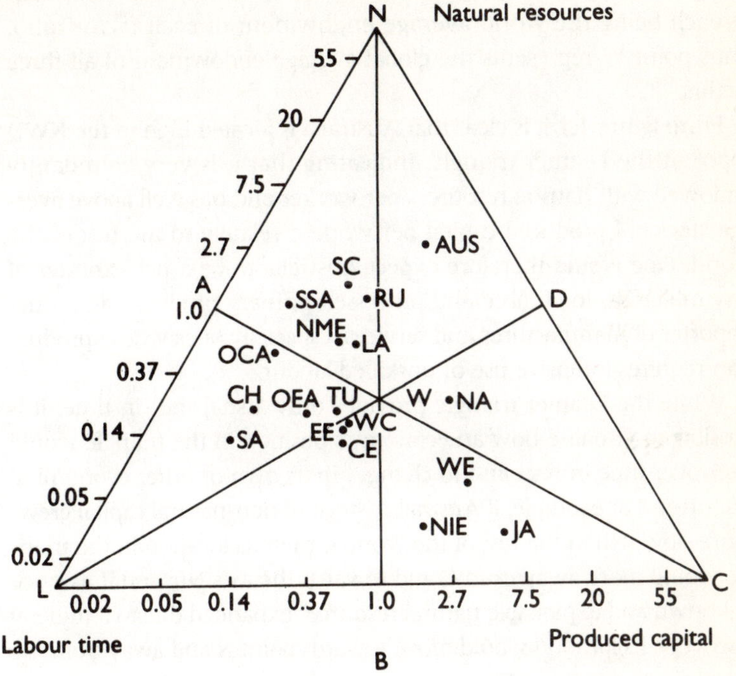

Note: The distance along LN from L measures the population density as a ratio of the world average (2.50 hectares *per capita*). The distance along LC from L measures gross national product *per capita* as a ratio of the world average ($4040). Both scales are in logs. Along any ray from C to the NL line the population density is constant, and similarly for rays from the other two corners of the triangle. W is the world's endowment point. Countries are represented as follows: AUS Australia; CE Czech, Hungary, Poland, Slovakia; CH China; EE Albania, Bulgaria, Romania, Yugoslavia; JA Japan; LA Latin America excluding Mexico; NA North America including Mexico; NIE Hong Kong, Singapore, South Korea, Taiwan; NME North Africa and Middle East; OCA Cambodia, Laos, Mongolia, Myanmar, North Korea, Vietnam; OEA Brunei, Indonesia, Philippines, Malaysia, Thailand, Pacific islands; RU Russia; SA South Asia; SC Armenia, Azerbaijan, Georgia, Kazakhstan, Kyrgyzstan, Tajikistan, Turkmenistan, Uzbekistan; SSA Sub-Saharan Africa; TU Turkey, Cyprus, Malta; WE Western Europe; WC Belarus, Estonia, Latvia, Lithuania, Moldova, Ukraine.

Source: adapted from Leamer (1987) using data from the World Bank (1993a).

etc.). Proxies used here to represent the natural resources to labour ratio and the produced capital to labour ratio are land area *per capita* and national product *per capita*.[1] These ratios are measured in log terms along the NL and LC sides of the triangle respectively, the mid-points of each being the world average endowment of each factor ratio. Thus point W represents the global average endowment of all three factors.

From figure 1.1 it is clear that Australia is located high in the NWD space of the Leamer triangle, indicating that it is very abundantly endowed with natural resources per worker and has well above average stocks of produced capital per worker, relative to the rest of the world. One would therefore expect Australia to be a net exporter of raw materials, food, fibre and processed primary products, and a net importer of manufactures and services, especially ones whose production requires intensive use of unskilled labour.

While the Leamer triangle provides only a snapshot in time, it is possible to visualise how an economy's position in the triangle would alter over time in response to changes in its own or other economies' resources. For example, if Australia's stock of non-natural capital grew more slowly than the rest of the world's, then its location in the triangle would more away from C and towards the axis NL; and if its stock of known and exploitable natural resources expanded due to a mineral discovery, its location would move towards point N and away from the base LC, *ceteris paribus*. The latter would also happen if there was a disappearance of natural resources in the rest of the world, which is effectively what happened when, in 1973, OPEC (the Organisation of Petroleum-Exporting Countries) formed its cartel and decided to restrict production and exports of petroleum. In that case, Australia's comparative advantage in and exports of energy raw materials increased (as did its comparative disadvantage in manufactures).

The Leamer triangle is not able without additional dimensions to distinguish between different types of natural resources. But it is not difficult to conceptualise how the country's comparative advantage in agricultural goods would fare if there was a new discovery of minerals in Australia or a depletion of available resource stocks abroad (and hence an increase in the international price) of energy raw materials: mobile resources such as labour and investable capital would move from non-mining sectors, including agriculture, and into the mining sector, boosting mining's shares of output, employment and exports.

With the ideas behind this triangle in mind, we can now examine the changing pattern of Australia's trade.

THE CHANGING COMPOSITION OF AUSTRALIA'S TRADE

Given Australia's relative abundance of natural resources per worker, it is not surprising that historically its exports have been overwhelmingly primary products. Apart from the gold rush periods, farmers and graziers supplied most of the country's export earnings up to the late 1950s (table 1.1). Then the opening up of mineral market opportunities in Japan stimulated exploration in the mining sector. That, coupled with the OPEC-induced rises in the price of energy raw materials in international markets around 1973–74 and 1979–80, resulted in a massive increase in the importance of minerals and metals in Australia's exports. In the 1950s they accounted for less than 10 per cent of total exports, but by the 1970s their share exceeded 30 per cent, and since the 1980s it has been more than 40 per cent. Meanwhile, agriculture's share of Australia's exports fell from more than 80 per cent in the 1950s to little more than 20 per cent of all goods and services now.

Table 1.1 Contribution to exports of goods and services by sector, Australia, 1950–51 to 1992–93 (%)

	Rural products	*Fuels, minerals & metals*	*Other merchandise*	*Services*	*TOTAL*
1950–51	86	6	3	5	100
1960–61	66	8	13	13	100
1970–71	44	28	12	16	100
1980–81	39	34	11	16	100
1985–86	32	42	9	17	100
1989–90	26	41	14	19	100
1992–93	23	41	16	20	100

Source: Australian Bureau of Agricultural and Resource Economics, *Commodity Statistical Bulletin 1990* and *Agricultural and Resources Quarterly*, December 1993, Canberra: AGPS.

Also notable from table 1.1 is that non-primary products comprise an increasing share of Australia's exports. Since the mid 1980s services have increased their share by a quarter, and the share of manufactures (other than processed primary products) has risen by half. Their combined share has thus risen in less than a decade from 26 to 36 per cent of all exports, to more than half as much again as agriculture's share. Moreover, within the main merchandise categories, higher value-added goods are dominating Australia's export growth and, according to recent survey reports, much of that growth has come from small- and medium-sized firms (Swan & Zeitsch 1992, McKinsey & Co. 1993).

The import composition of Australia's trade has been considerably more stable than the export composition, but it is clear from table 1.2 that manufactures dominate imports and primary products are of minor significance.

The net trade picture is summarised in table 1.3. There each sector's net exports (gross exports minus gross imports) is expressed as a ratio of the sum of sectoral exports and imports. This indicator of trade specialisation therefore takes a value between minus one and plus one. It has fluctuated around plus two-thirds but shown no trend since 1950 for primary products. For manufactures this indicator is highly negative but has been declining, suggesting that Australia is becoming more involved in intra-industry trade in manufactures and gaining competitiveness in some niche markets. And for services this indicator is neg-

Table 1.2 Composition of Australia's imports of goods and services, 1950–51 to 1992–93 (per cent)

	Primary products	*Machinery*	*Transport equipment*	*Other merchandise*	*Services*	*TOTAL*
1950–51	22	14	8	40	16	100
1960–61	18	17	7	37	21	100
1970–71	11	20	10	34	25	100
1980–81	18	20	9	30	23	100
1984–85	12	22	9	35	22	100
1992–93	11	22	11	33	23	100

Source: Australian Bureau of Statistics, *Composition of Trade*, Canberra: AGPS, various issues; and Reserve Bank of Australia, *Australian Economic Statistics, 1949–50 to 1984–85*, Sydney: RBA.

Table 1.3 Index of trade specialisation, Australia, 1950–51 to 1992–93 (current value of net exports/exports plus imports)

	Primary products	*Manufactures*	*Services*
1950–51	0.67	–0.89	–0.45
1960–61	0.55	–0.70	–0.32
1970–71	0.73	–0.69	–0.23
1980–81	0.57	–0.72	–0.24
1990–91	0.66	–0.66	–0.16

Source: see Tables 1.1 and 1.2.

ative but not highly so, especially in more recent years as services exports (notably tourism and education services) have risen more rapidly than services imports.

A more precise indication of Australia's comparative advantage *vis-à-vis* other countries, as revealed by these trade data, can be found by comparing the sectoral shares of table 1.1 for Australia with those for the rest of the world (Balassa 1965). In 1991, for example, the value of world trade in goods and services comprised 10 per cent agricultural goods, 12 per cent fuels, minerals and metals, 57 per cent other merchandise, and 21 per cent commercial services (GATT 1992). Australia's shares for primary exports (imports) are clearly well above (below) those for the rest of the world and conversely for manufactures. The contrast is even more marked between Australia and other high-income countries: for North America, Western Europe and Japan, primary products in 1991 accounted for only 19, 15 and 3 per cent of their total exports respectively, while manufactures accounted for 58, 62 and 84 per cent respectively (the residuals being services). With respect to services, Australia's imports slightly exceed its exports, but the gap has been narrowing, and in the early 1990s both shares have been close to the 1991 global average of 21 per cent (and to the Western European and North American averages of 22 and 23 per cent respectively). This suggests the country has had a slight comparative disadvantage in services trade but that, if present trends continue, it might soon switch to having a slight comparative advantage in that product group, at least as revealed by these trade data.

THE CHANGING DIRECTION OF AUSTRALIA'S TRADE

Accompanying these changes in the composition of trade have been equally dramatic changes in the direction of Australia's commerce. Until the 1950s more than 60 per cent of Australia's international trade and investment was with Western Europe, dominated of course by the United Kingdom. Then during the 1960s America's importance doubled and Japan became a major trading partner. That was followed by a steady increase in the importance of a growing number of developing countries of North-east and South-east Asia. By the early 1990s Europe accounted for only a fifth of Australia's merchandise trade compared with more than three-fifths in the 1950s, during which time the share of the Asia-Pacific Economic Cooperation (APEC)[2] region had grown from a quarter to nearly three-quarters (table 1.4). Yet over that period Europe's share of world trade has remained around 50 per cent, and the APEC region's percentage share has increased only from the low thirties to the high thirties. Clearly, Australia's trade with APEC countries has grown far faster than the region's importance in world trade: the intensity of that trade has gone from weak to strong in a relatively small number of years.

Accompanying this shift in the direction of merchandise trade have been equally dramatic redirections of Australia's services trade and investment. The sum of services exports and imports with APEC economies increased from around 50 to nearly 60 per cent of Australia's services trade during the 1980s, and the Australian dollar value of the stock of foreign investments both by Australians in other APEC economies and by APEC members in Australia increased six-fold over that decade (figure 1.2).

INFLUENCES OF POLICIES ON AUSTRALIA'S TRADE PATTERN

The above changes in the composition and direction of Australia's trade have been influenced not only by changes in the standard determinants of comparative advantage, such as relative resource endowments, but also by policy changes at home and abroad. Details of pertinent

Table 1.4 Direction of Australia's merchandise trade, 1951–92 (%)

	1951–55	*1968–72*	*1980–84*	*1990–92*
Exports				
United Kingdom	36	11	4	4
Other Europe	27	16	10	12
North America	10	16	13	12
Japan	8	26	27	26
New Zealand	4	5	5	5
Developing East Asia	6	12	20	30
Middle East	1	2	8	3
Other developing	8	12	13	8
TOTAL	**100**	**100**	**100**	**100**
(of which **APEC**	**28**	**49**	**65**	**73**)
Imports				
United Kingdom	45	21	7	6
Other Europe	15	19	17	20
North America	15	27	25	26
Japan	2	13	21	18
New Zealand	1	2	3	5
Developing East Asia	7	7	14	19
Middle East	4	5	9	3
Other developing	11	6	4	3
TOTAL	**100**	**100**	**100**	**100**
(of which **APEC**	**25**	**40**	**63**	**68**)

Note: APEC comprises all the major trading economies of North America, East Asia and Australasia.

Source: Australian Bureau of Statistics, *Overseas Trade*, various issues.

Figure 1.2 Australia's services trade and foreign investment with APEC economies, 1981–82 to 1991–92

(a) APEC's share of Australia's trade in services (exports plus imports)

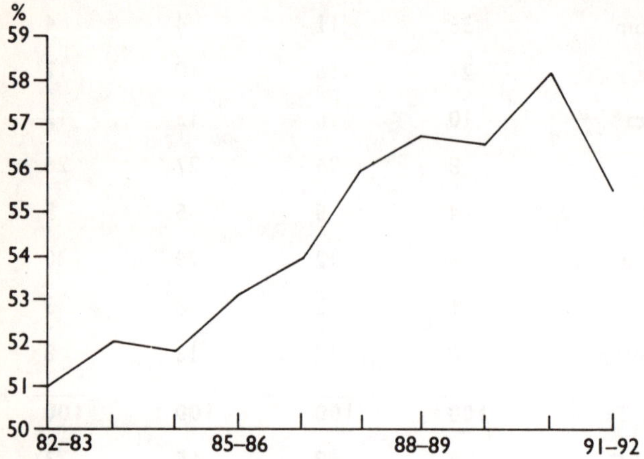

(b) Australia's investments in other APEC economies (black) and their foreign investments in Australia (white), in current Australian dollars

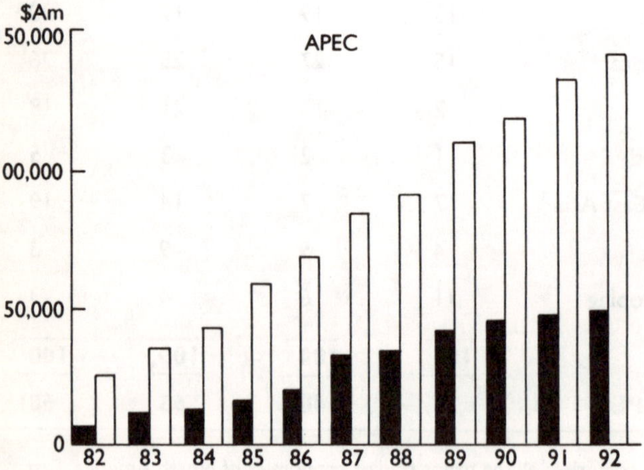

Source: Department of Foreign Affairs and Trade (1993).

policy changes are provided in subsequent chapters, but a few need to be mentioned here before assessing Australia's economic growth performance. Overseas, a key reason for the decline in agriculture's share of Australia's exports has been the growth in agricultural protectionism in industrial countries. The increasing farm price support provided by the European Community's Common Agricultural Policy since the late 1950s, and the United Kingdom's accession to the EC in 1973 (and hence the replacement of Commonwealth preferential trade treatment by the UK towards Australian farm exports with ever-higher agricultural trade barriers) has been a double whammy: it reduced Australia's exports to Europe (table 1.4), and it ensured that a lower price would be received for farm exports in other markets. This protection growth contributed significantly to the decline in the international price for farm relative to manufactured goods, and to the fluctuations around that downward trend (Anderson & Hayami 1986; Tyers & Anderson 1992), the extent of which is shown in figure 1.3.

Figure 1.3 Real international food prices, 1900–92[1] (1977–79 = 100)

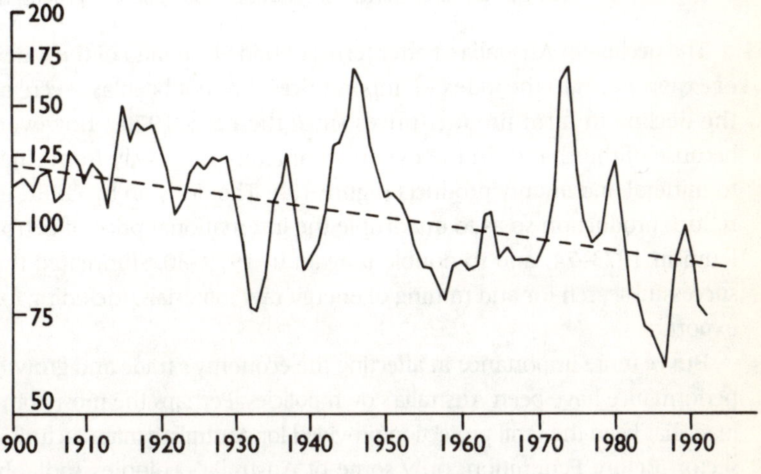

1. An index of export prices in US dollars for cereals, meats, dairy products and sugar, deflated by the US producer price index (primarily for industrial products), with weights based on the importance of each product in global exports, 1977–79.

Source: Author's calculations based mainly on price series from the World Bank's Economic Analysis and Projections Department (see Grilli & Yang 1988).

Figure 1.4 Australia's barter terms of trade, 1974–93 (1989–90 = 100)

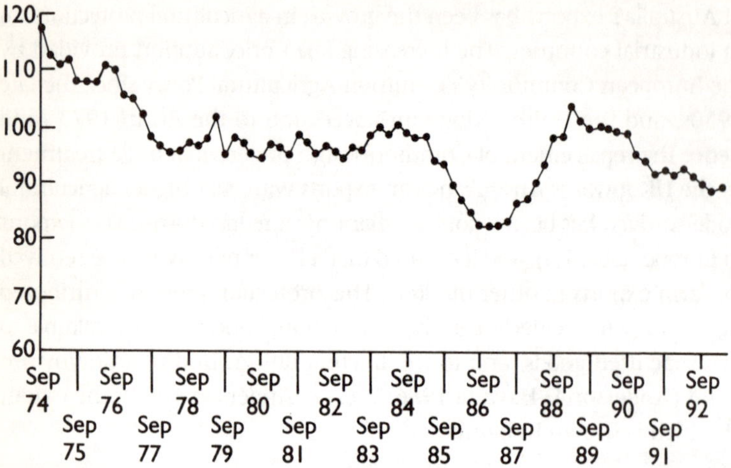

Source: Australian Bureau of Statistics (1993), *Balance of Payments*.

The decline in Australia's barter terms of trade (the ratio of the index of export prices to the index of import prices) has not been as severe as the decline in farm product prices since the early 1970s, however, because of the change in the export commodity mix away from rural to mineral and energy products (figure 1.4). The decision by OPEC to reduce production so as to quadruple the international price of petroleum in 1973–74, and to double it again in 1979–80, stimulated the successful search for and mining of energy raw materials, including for export.

But of more importance in affecting the economy's trade and growth performance have been Australia's own policies. Perhaps the most dominant has been the tariff protection provided for Australia's manufacturing sector. Before Federation, only some of Australia's colonies (notably Victoria) had high tariff walls, but soon after Federation in 1901 a national policy of tariff protection against imports of manufactures was adopted. Initially the average rate of duty was modest relative to that of other industrial countries (table 1.5), but at that time Australian manufacturers also enjoyed a very considerable degree of natural pro-

Table 1.5 Average manufacturing tariff rates, various industrial countries, 1902, 1913, 1925 and 1970 (%)

	1902	*1913*	*1925*	*1970*
AUSTRALIA	6	16	27	23
Belgium	13	9	15	na
Canada	17	26	23	14
Denmark	18	14	10	na
EEC-6[1]	na	na	na	8
France	34	20	21	na
Germany	25	13	20	na
Italy	27	18	22	na
Japan	10	20	13	12
Netherlands	3	4	6	na
New Zealand	9	na	na	23
Norway	12	na	na	11
Sweden	23	20	16	7
Switzerland	7	9	14	3
United States	73	44	37	9

na = Not available
1 Belgium, France, West Germany, Italy, Luxembourg and the Netherlands.

Source: Anderson and Garnaut (1987, table 2.1), based on the League of Nations, *Tariff Level Indices*, Geneva (1927), and the GATT, *Basic Documentation of the Tariff*, Geneva (1972).

tection in the form of high ocean transport costs facing Northern Hemisphere competitors. During the subsequent seven decades, other industrial countries gradually lowered their tariffs on manufactures while Australia progressively raised many of its tariffs. By 1970 only New Zealand had an average tariff rate as high as Australia's (table 1.5).

An important consequence of this policy choice is that Australia moved from being in 1870 well above average in its trade to GDP ratio (taking into account the size of its economy) to being well below average during the post-World War II decades. As figure 1.5 shows, the drop in its ratio of exports plus imports to GDP was from around 0.6 to around 0.3. That ratio has certainly grown since the mid 1970s, but

Figure 1.5 Trade orientation and the GDP of various industrial countries, circa 1870 and 1984[1]

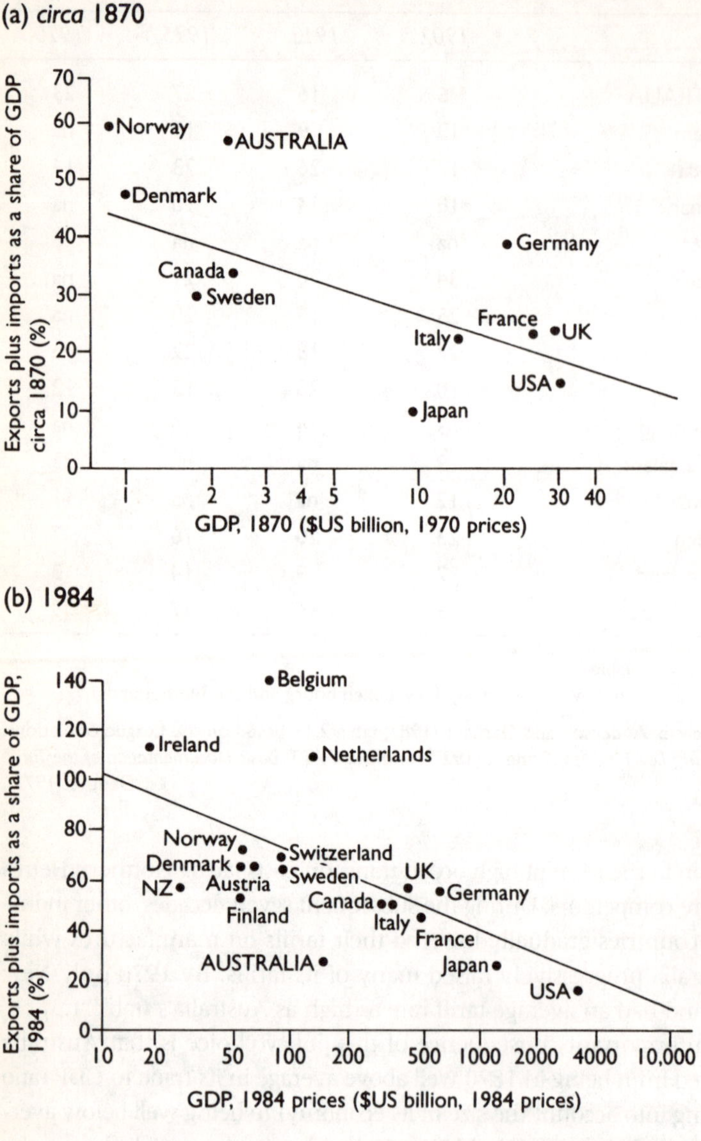

1 The least-squares regression line was estimated excluding Australia.

Source: Anderson and Garnaut (1987, figures 2.3 and 2.4).

Table 1.6 Trade orientation of Australia and other OECD countries, 1963–91 (volume of exports plus imports as a percentage of real GDP at 1989–90 prices for Australia and 1985 prices for other countries)

	1963	*1975*	*1987*	*1991*	*Growth (% pa) 1963–91*
Australia	24	20	30	37	1.4
Canada	30	40	56	71	2.5
France	25	39	49	56	2.7
Germany	31	49	61	75	2.7
Ireland	58	80	129	136	3.0
Japan	14	23	24	27	2.2
Netherlands	66	97	121	134	2.3
New Zealand	50	54	71	80	1.6
Switzerland	43	54	80	82	2.6
United Kingdom	36	48	59	64	2.0

Source: Industry Commission (1993, p. 103)

not as fast for Australia as for other OECD countries and from a much lower base (table 1.6).

Eventually this inward-looking development strategy, which was gradually perceived from the early 1970s as having contributed to Australia's relatively poor economic growth performance (see next section), began to change. In 1973 the government cut all tariffs by 25 per cent, and in the subsequent two decades tariffs were reduced further and further. True, two industry groups (motor vehicles and parts, and textiles, clothing and footwear) managed to persuade the government to give them extra protection in the form of import quotas for a time, but since 1985 even that protection has been phased down. As a consequence, the effective rate of protection to manufacturing has been more than halved since the early 1970s (bringing it down to that for agriculture — see figure 1.6a), and it is scheduled to be reduced by a further two-thirds during the 1990s (see figure 1.6b), which will bring it close to the average for other industrial countries. An indication of the extent of the trade consequence of this liberalisation is shown in figure 1.7: when measured at constant prices (to net out the effect of the

Figure 1.6 Effective rates of assistance to Australia's manufacturing and agricultural sectors, 1968 to 1992 with projections to 2000

(a) Total manufacturing and agriculture

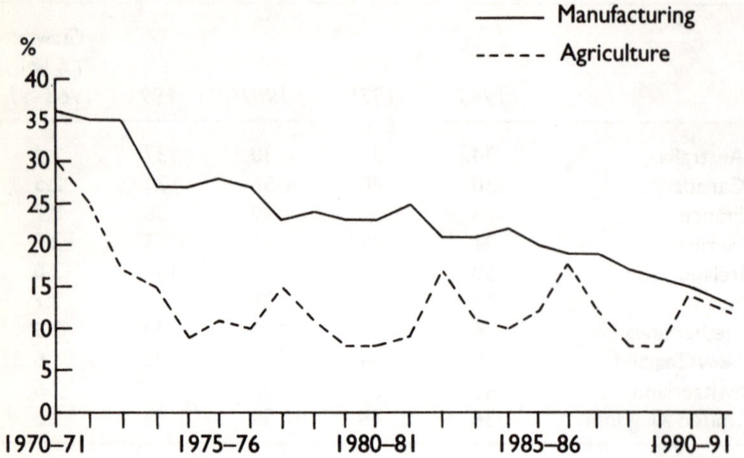

(b) Textiles, clothing and footwear (TCF), motor vehicles and parts (MVP) and other manufacturing industries

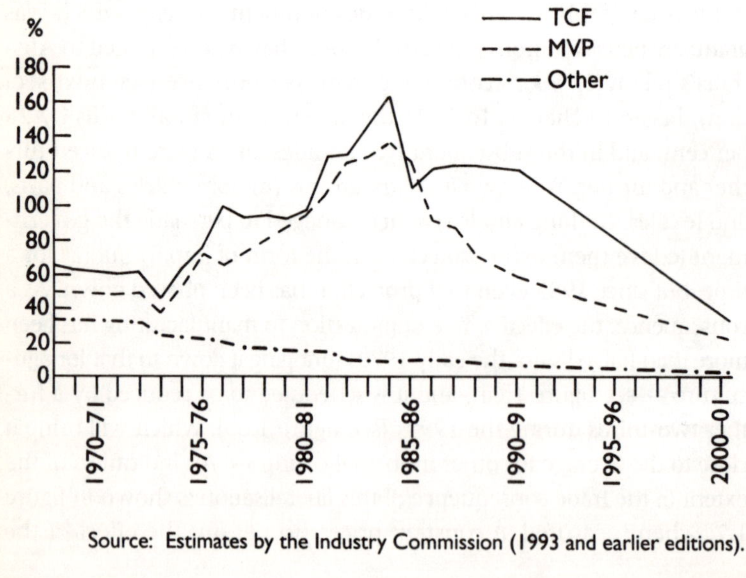

Source: Estimates by the Industry Commission (1993 and earlier editions).

deterioration in the terms of trade shown in figure 1.4), the ratios of exports and imports of goods and non-factor services to GDP have grown by 50 per cent since the early 1970s when the policy reform began.

AUSTRALIA'S ECONOMIC GROWTH PERFORMANCE AND STRUCTURAL CHANGES

During the latter part of the nineteenth century and early this century, the Australian population enjoyed perhaps the highest measured income *per capita* in the world (Maddison 1982). But by 1920 it was second to the US and by 1950 it was third to Canada, and since then it has been overtaken by numerous other countries. In 1991 it was ranked eighteenth at current prices converted at official exchange rates (table 1.7), and virtually the same using purchasing power comparisons (World Bank 1993b, p. 297).[3]

Figure 1.7 Share of GDP traded as goods and non-factor services, at constant prices, Australia, 1971–91 (based on 1987 Australian dollars)

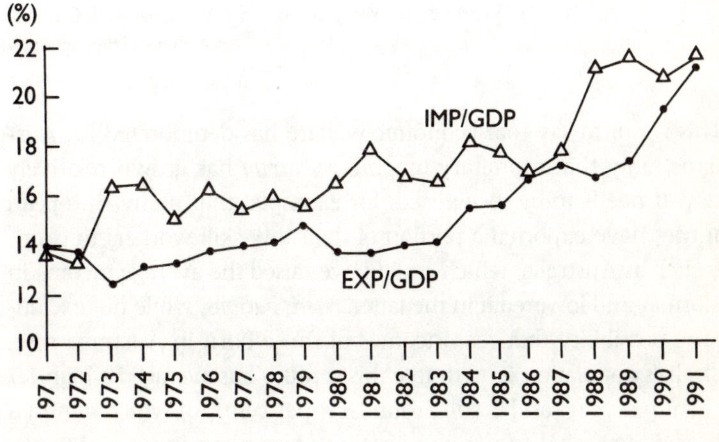

Source: World Bank (1993a, pp. 96–7).

Table 1.7 Ranking of high-income countries by GDP *per capita*, 1900–91[1]

1900	*1950*	*1970*	*1991*
1 AUSTRALIA (?)	1 United States	1 United States	1 Switzerland
	2 Canada	2 Sweden	2 Luxembourg
	3 AUSTRALIA	3 Canada	3 Japan
		4 Switzerland	4 Sweden
		5 Denmark	5 Norway
		6 Germany, France	6 Finland
		7 AUSTRALIA	7 Denmark
			8 Germany
			9 Iceland
			10 United States
			11 Canada
			12 France
			13 Austria
			14 United Arab Emirates
			15 Belgium
			16 Netherlands
			17 Italy
			18 AUSTRALIA

1 At current prices converted to US dollars at official exchange rates.

Source: World Bank, *World Tables 1993*, Washington, DC (1993) and earlier editions.

This is not to say that economic welfare has deteriorated for Australians, only that monetary income *per capita* has grown relatively slowly. It needs to be recognised, for example, that many European countries have exported a portion of their low-skill workers to countries such as Australia, which could have raised the average income in the former and lowered it in the latter, *ceteris paribus*, while not excluding the possibility that real incomes of those born in Australia rose. Australia's population growth rate this century has averaged a high 1.7 per cent per year (and is still around 1.4 per cent), of which almost a third is directly due to net immigration (Australian Bureau of Statistics 1992). None the less, as mentioned above, an increasing proportion of voters have perceived that Australia's trade and other microeco-

Table 1.8 Growth in total and *per capita* GDP, various countries and country groups, 1960–90 (% per year)

	1960–68	*1968–73*	*1973–79*	*1979–90*	*1960–90*
Total GDP					
Australia	5.0	5.4	2.8	3.1	3.9
Austria	4.2	5.9	2.9	2.3	3.5
Canada	5.5	5.4	4.2	2.8	4.2
European Union	4.7	4.9	2.5	2.2	3.4
Japan	10.2	8.6	3.6	4.1	6.3
New Zealand	3.3	5.0	0.7	1.5	2.4
Norway	4.4	4.1	4.9	2.6	3.8
Sweden	4.4	3.7	1.8	1.9	2.8
United States	4.5	3.2	2.4	2.6	3.2
ALL OECD	**5.1**	**4.6**	**2.7**	**2.7**	**3.6**
***Per capita* GDP**					
Australia	3.0	3.5	1.5	1.6	2.3
Austria	3.6	5.4	3.0	2.1	3.2
Canada	3.6	4.1	2.9	1.8	2.9
European Union	3.8	4.2	2.1	1.9	2.8
Japan	9.1	7.1	2.5	3.5	5.3
New Zealand	1.4	3.4	–0.2	0.8	1.2
Norway	3.6	3.3	4.4	2.2	3.2
Sweden	3.6	3.1	1.5	1.6	2.4
United States	3.1	2.0	1.4	1.6	2.0
ALL OECD	**3.9**	**3.5**	**1.9**	**1.9**	**2.7**

Source: OECD, *Historical Statistics 1960–1990*, Paris: OECD, 1992

nomic policies contributed significantly to reducing the efficiency of resource use and to holding back income growth unnecessarily. As table 1.8 shows, despite the huge windfall in terms of the minerals and energy export boom, real *per capita* GDP grew at less than the OECD average between 1960 and 1990, and even total GDP growth was barely above the OECD average.

But Australia's protectionist trade policy certainly achieved one of its other objectives, which was to boost the manufacturing sector. In these times when industrial pollution is considered undesirable it is difficult to imagine why people considered industrial growth in particular to be something worth fostering, but in earlier times they did, not

Table 1.9 Changing structure of the Australian economy, 1900–92

	Agriculture	*Mining*	*Manufacturing*	*Other*
Share (%) of GDP at factor cost				
1900–01	19.3	10.3	12.1	58.3
1913–14	23.5	5.1	13.4	58.0
1928–29	21.2	1.8	16.7	60.3
1938–39	19.5	3.3	18.5	58.7
1948–49	21.3	2.5	26.2	50.0
1955–56	15.9	2.3	28.0	53.8
1962–63	12.6	1.7	26.8	58.9
1968–69	9.6	2.4	26.1	61.9
1973–74	9.6	4.0	23.2	63.2
1980–81	5.4	6.5	20.6	67.5
1986–87	3.8	5.4	16.8	74.0
1991–92	3.1	4.4	15.3	77.2
Share (%) of employment				
1962–63	10	1	26	63
1972–73	7	1	24	68
1982–83	6	1	19	74
1991–92	5	1	15	79

Note: The adoption in 1971 of a new Australian standard industrial classification means that the series from 1962–63 are not entirely consistent with the previous years.

Sources: Boehm (1979, table 1.1); Australian Bureau of Statistics, *Australian National Accounts, National Income and Expenditure*, and *The Labour Force*, Canberra, various issues.

least because it was feared that the economy would otherwise consist of no more than quarries and farms. At the start of the century, less than an eighth of GDP was contributed by manufacturers, but with protection increases their share rose steadily during the next sixty years. It peaked at about 30 per cent around 1960, before the mining boom and other deindustrialisation forces caused it to decline (table 1.9). At its peak Australia's manufacturing share of GDP equalled the average for advanced industrial countries and the sector's employment share exceeded the OECD average (columns 1 and 4 of table 1.10), despite the fact that the country's relative resource endowment ratios and trade composition suggest that it has a strong comparative disadvantage in manufacturing.

Table 1.10 Share of manufacturing in GDP and employment, various countries and country groups, 1960, 1974 and 1990 (percentages)

	Share of GDP			*Share of employment*		
	1960	*1974*	*1990*	*1960*	*1974*	*1990*
Australia	29	25	15	31	25	17
Austria	35	31	26	31	32	27
Canada	23	20	17	25	22	16
European Union	33	30	23	29	30	23
Japan	35	34	29	21	27	24
New Zealand	na	24	17	26	26	17
Norway	21	22	14	26	24	16
Sweden	na	27	20	32	28	21
United States	28	23	18	26	24	18
ALL OECD	**29**	**27**	**22**	**26**	**27**	**21**

Source: OECD, *Historical Statistics 1960–1990*, Paris: OECD (1992).

Since the tariff reductions began, however, the manufacturing sector's importance in the economy has shrunk much more for Australia than for other industrial countries, bringing it down well below the OECD average (table 1.10). The feared reduction of the economy to farms and quarries has proved unwarranted, with the share of employment in services actually increasing, as has the extent of urbanisation. In Australia today more than two-thirds of all jobs are in the service sector (six percentage points more than the OECD average), and all but a seventh of the population lives in urban areas (compared with the OECD average of three-quarters).

While a relatively smaller manufacturing sector is certainly more consistent with Australia's comparative advantages (and its importance will be even smaller if/when the announced remaining tariff cuts for the rest of the 1990s are implemented), more structural adjustment within the manufacturing sector and further improvement in economic performance are still possible. One set of indications of structural adjustment and improved economic performance are provided in table 1.11, which shows the changing extent to which domestic sales of manufactures are supplied by imports and domestic production of manufactures are exported. In the absence of protection policies one would expect Australia's manufacturing import penetration to be above average and its

Table 1.11 Import penetration and export propensity of manufacturing, various industrial countries, 1971–90

	Import penetration			*Export propensity*		
	Growth rate		*Level*	*Growth rate*		*Level*
	1971–80	*1981–90*	*1985–90*	*1971–80*	*1981–90*	*1985–90*
Australia	4.7	2.2	24	2.5	–0.7	13
Canada	2.2	2.5	36	2.5	1.5	35
Finland	0.0	2.3	30	2.7	0.4	34
France	4.0	3.4	28	4.5	1.9	27
Germany (West)	4.2	2.6	25	3.4	2.4	32
Italy	3.0	0.8	21	3.8	–1.1	23
Japan	3.7	0.8	5.4	2.8	–2.3	12
Netherlands	2.7	2.8	65	2.5	2.0	67
Norway	–0.6	0.8	42	–0.5	0.3	31
Sweden	2.5	1.2	41	2.5	0.8	45
United Kingdom	5.1	3.3	29	4.5	1.3	25
United States	6.1	6.5	13	6.9	1.3	8
UNWEIGHTED **Average**	**3.1**	**2.4**	**30**	**3.2**	**0.7**	**30**

Notes: *Import penetration* is defined as imports as a percentage of domestic sales (= domestic production minus exports plus imports); *export propensity* is exports as a percentage of domestic production. Growth rates are the average annual percentage increase over the decade shown.

Source: Oliveira Martins (1993), based on the OECD's DSTI STAN database.

export propensity to be below average; and, given its dramatic tariff cuts since the early 1970s relative to those of other industrial countries, one would expect both its import penetration ratio and its export propensity to have grown faster than average. In fact the manufacturing import penetration ratio for Australia grew above the OECD average in the 1970s but a little below the average in the 1980s so that in the latter 1980s its level was still well below instead of above average (24 compared with 30 per cent). And Australia's propensity to export manufactures remains very low, having grown much more slowly than the OECD average despite the reduction in incentives to employ resources in the import-competing subsector of manufacturing.[4]

QUESTIONS ARISING FROM THIS REVIEW

The above facts raise numerous questions, some of which are the focus of subsequent chapters. The following are mentioned by way of illustration. A fundamental one is: how important is trade, and hence policies that affect trade, for growth of incomes in small open economies such as Australia's? In particular, is rapid income growth possible following trade policy reform without substantial deregulation of labour markets and other microeconomic reforms, and how long will it take before incomes grow noticeably faster?[5] Another is: why did Australia adopt a highly protectionist approach towards its manufacturing sector (as did two other high-income, natural-resource-rich countries whose economies have grown even slower than Australia's, namely Argentina and New Zealand)? Why too did Australia not reduce its tariffs on manufactures (and raise farm price supports) like other OECD countries in the quarter century following World War II, but then begin reforming later and continue the reform program despite a severe recession? Why were textiles and cars exempted from reform for another decade, but then also subjected to major cuts in protection? Turning to the location of jobs and people, why are such large shares of jobs in the service sector and people located in cities, despite the tariff reform and a boom in mining? What roles have immigration and foreign capital inflows played (*vis-à-vis* domestic population growth and national savings) in affecting the country's relative resource endowments and income growth rate, and how have they been influenced by our trade policies? How have other micro- and macro-economic policies impacted on Australia's trade and growth performance? And to what extent will growth be affected by the successful conclusion to the Uruguay Round of GATT multilateral trade negotiations and the spread of regional integration agreements? These and related questions are addressed in the chapters to follow.

FURTHER READING

A documentary history of Australia's trade and sectoral policies and their effect on the structure and growth of the economy can be found in the annual reports of the Industry Commission (formerly called the

Industries Assistance Commission and, before 1973, the Tariff Board). Anderson and Garnaut (1987) provide details of the extent, causes and effects of Australia's protectionism to the early 1980s, and the subsequent decade of reform is covered comprehensively in Garnaut (1994). Australia's recent and prospective growth performance is discussed in more detail in Howe (1993). Details of other microeconomic reforms can be found in Forsyth (1992). Driving the reforms was an ascension of 'economic rationalism' in bureaucratic and media debates, discussed in King and Lloyd (1993).

2

EVOLUTION AND POLITICAL ECONOMY OF AUSTRALIAN TRADE POLICIES

Jonathan Pincus

Since the mid nineteenth century, Australian trade policy has been a central component of a package of measures used by governments to redistribute the benefits of abundant Australian natural resources. Australian trade policy since the 1850s has been set within a wider policy framework, that of stimulating the growth and shaping the pattern of development of the economy. Therefore, we need to consider the economic linkages that existed between trade policy, and wages policies, immigration policies and taxation and spending policies.

Although trade policy was not designed solely by nor implemented single-mindedly by 'benevolent dictators' seeking to maximise the economic welfare of the Australian residents, neither was trade policy merely the plaything of selfish, rent-seeking pressure groups. Rather, it was shaped by both sorts of interests, by the 'public interest' as well as by 'private interests'; it was a product of conceptions of what is good for the nation, as well as of understandings of what is good for 'me and my mates'. Australian trade policy has been made within a political system, in which there are foreign as well as domestic pressures of ideas, opportunities and interests.

Until the 1970s the prevailing view had been that the Australian economy had too few people and too small a productive capacity, so that it offered too few jobs even at full employment. A larger population of the right (that is, white) kind and a bigger economy were expected to bring widespread economic benefits in the form of increased opportunities for reaping economies of scale and improvements in the capacity to mount a defence of the nation and its allies. Trade policy measures, especially in the form of assistance to import-competing industries through tariffs, in effect were used to attract immigrants and jobs to Australia. The strategy of stimulating an expansion in the size of the Australian population and economy, even if this was at the expense of a lower standard of living than would otherwise obtain, was widely regarded as successful, especially during the 'long boom' of the 1950s and 1960s. Increasingly in recent decades, however, both the developmental objective of securing and retaining a larger population, as well as the means used — an inward-looking trade policy — have been thrown into serious doubt. Protectionism fell into disrepute. The balance has changed within and between private interests and public interests; elite opinion is now predominantly behind a movement towards a more open, outward-looking, internationally competitive Australian economy.

The remainder of this chapter is a summary of the history of Australian trade policy, which becomes more sketchy as we get closer to the present: policies of recent decades are discussed in greater detail in later chapters. Table 2.1 provides a brief listing of the major events in the history of Australian trade policy.

PROTECTIONISM AFTER THE GOLD RUSHES OF THE 1850S

Governments in the colony of Victoria after the gold rushes turned to tariffs and other trade-distorting policy instruments in an attempt to stimulate employment opportunities. Thus an abiding theme that runs through our story until recent decades, which is that a chief objective of tariffs and other trade-distorting instruments was to create jobs so as to attract and retain population, began in the 1860s.

Before attempting to explain why the other major colony, New South Wales, did not follow the Victorian lead with any lasting enthusiasm, it is useful first to define *import protection*. A tax on an import provides

Table 2.1 Major events in Australian trade policy, 1800–1991

Years	*Event*
1800	First import duties
1802	Customs tax preference to British goods
1819–22	'Old Colonial System' imposed on NSW
1843	Intercolonial tariff preferences denied
1850	Australian Colonies Government Act
1863	Failure of Intercolonial Conference
1866	Victorian tariff imposed
1873	Australian Customs Duties Act allows intercolonial preferences
1879	Victorian tariff rates increased
1885	South Australian 'protective' tariffs
1889	Victorian rates raised
1891	Higher NSW rates
1895	Victorian rates raised; income taxation imposed
1895	NSW lowers rates; income taxation imposed
1902	First federal tariff enacted
1906	'New Protection' declared unconstitutional
1908	Lyne tariff; preference to Britain
1913	Interstate Commission begins hearings
1914	Extension of preference to Britain
1919	Britain grants some imperial preferences
1921	Greene tariff: higher rates; deferred duties; British preferences; anti-dumping strengthened. Tariff Board established
1926	Pratten tariff
1929	Brigden Report on the tariff
1929–31	Scullin emergency duties and other measures
1932	Ottawa Agreement
1933	Exchange Adjustment Act: 25 per cent to British Preferential Rates
1947	General Agreement on Tariffs and Trade
1952–60	Import quotas
1962	Special Advisory Authority for temporary assistance
1965	Vernon Report
1964–67	Kennedy Round of GATT; Australia abstains
1967	Rattigan announces Tariff Board systematic review
1973	Across-the-board 25 per cent tariff cut
1974	Industries Assistance Commission and Temporary Assistance Authority established
1974–79	Tokyo Round of GATT; Australia mostly abstains
1975	Manufacturing Green Paper (Jackson Report)
1977	Manufacturing White Paper
1979	Study Group on Structural Adjustment (Crawford Report)
1983	Button's steel plan announced
1983	Australian dollar floated
1987	IAC completes a general review; IAC budget cut
1987	May economic statement
1989	Garnaut Report
1991	Most tariffs to be cut to 5 per cent by 1996; motor vehicles and textiles, clothing and footwear exempted

protection to the local industry if no excise tax is levied on the production or sale of the import substitute or if excise is levied at a rate lower than the tariff;[1] thus protection (or industry assistance) is absent if tax rates are uniform. Now, revenue considerations cause tax rates to vary among products according to elasticities of demand and supply, generating what is called 'incidental protection'. The Victorian tariffs were considered deliberately protective in that the tax rates varied not only for reasons of tax revenue but also according to employment or defence considerations or lobbying skill.

Revenue was a prime concern. Colonial governments had various ways to garner revenue from the major asset that the British ceded them, namely, the land and its contents. In turn, the revenue was used to subsidise immigrants and to service the public debt incurred for works and other purposes, some of which increased the value of the public revenue base. There were three main sources of public revenue: export taxes; import taxes; and land sales, leases or taxes. Both Victoria and New South Wales imposed taxation on gold exports. Both colonies used tariffs (as did other colonies) but to different extents. Surprisingly to most non-economists, the incidence (that is, who actually bears the economic cost) of a tax on exports is identical to the incidence of a tax on imports: where some factors of production are internationally mobile and others are not, both taxes fall ultimately on the immobile factors of production. Labour was fairly mobile in the nineteenth century, as evidenced by the net inflow of almost 300,000 people between 1851 and 1860 into Victoria, a flow more than four times its 1851 population. Victorian opinion leaders, like David Syme of the *Age* newspaper, feared that those who rushed in could exit quickly, for fresh gold fields or more settled occupations in other colonies. At that time, capital equipment was also relatively easily moved from one location to another.

The ultimate fixed factor was land and the structures and improvements on it. Therefore, taxes on external trade were alternatives to charges on land. When land and other revenues lagged, and the budget fell into deficits, the governments of the day resorted to higher taxation, including import taxation. Except for alcoholic spirits, it was then generally too costly to collect excise taxes (that is, taxes based on the quantity or value of local production). As economic conditions improved, and public revenues increased, the rates of tariff tended to be reduced somewhat. However, despite these reversals, Victoria was firmly protectionist. Patterson (1968:33–4) attributes the triumph of protectionism not only to the rhetoric of Syme, nor solely to revenue hunger, but also to the industrial composition of Victoria in the 1860s (and the protection itself would have reinforced this factor): the abundance, relative to New South Wales, of men whose interests would gain from protection of manufacturing. In contrast, New South Wales had a larger, steadier flow of land revenues and favoured lower, more uniform tariff rates; indeed, the free-trade radicals found it possible to force the issue, so that by the middle of the 1890s New South Wales had taxes for revenue only.

There is a continuing controversy among economic historians over the appropriate judgement to make about this Victorian colonial experiment. The balance of opinion is that the Victorian tariff policy probably had more effect on the composition of economic activity in Victoria before Federation and rather less on the aggregate size of the Victorian labour force and population. Although protection encouraged urban activities, it did so substantially at the expense of rural activities, which were then almost as labour-intensive (Siriwardana 1991). The rates of protection were neither very high nor permanent. Victoria suffered from tax evasion through smuggling, especially from South Australia, which placed some limits on the relative height of the Victorian tariff. Also, the structure of the Victorian tariff was not well designed to focus the protective possibilities in that it lacked strong 'escalation', that is, rates on semi-finished and intermediate inputs were not kept sufficiently low. There being no legal barriers on white migration from one colony to another, it is possible that whatever gains Victoria made in aggregate population from its policy of tariff protection, they were partly at the expense of population growth in the other colonies.

Federation and the Defeat of Free-Traders: 'New Protection'

The British Government, when legislating the framework for responsible colonial government in 1850, constrained the Australian colonies from imposing taxation that discriminated in favour of other colonies or parts of the empire. Therefore, the various independent colonies found themselves taxing the produce of other colonies, a matter of some regret and, later, friction. Thus, one of the chief economic objectives of Federation in 1901 was the creation of a single, Australia-wide market by integrating the separate colonial markets.

The new central government was given power over currency, coinage and legal tender; over banking, insurance, quarantine, corporations, bankruptcy, copyrights, weights and measures, posts and telegraph and the like; that is, over the ways in which the colonies had imposed non-tariff barriers on intercolonial trade. In addition, the federal constitution provided for the removal of discrimination in railway freight rates and some other, but not all, means of protecting one set of Australian producers from competition from other Australian producers. More important, and central to the achievement of this wider market, was the elimination of tariffs or customs duties, and the associated excises and bounties, that the various colonies had used to assist their own producers over those in other colonies. That is, Federation brought an Australian customs union with free trade between the states and a common external tariff against foreign products.

At what level and for what purposes would that common tariff be set? Revenue was a pressing, even a dominant, consideration. Customs duties yielded more than half the total Australian tax collections in each year until World War I and remained the most important source of public revenue until 1941, during World War II, when income taxes surpassed them. (Table 2.2 shows some data on the relative importance of import taxation.) Federation had given exclusive access to the largest tax revenue sources, customs duties and excise taxes to the nascent Commonwealth Government in 1901. For the first ten years, however, the Federal Parliament had to pass back to the states three-quarters of the revenue collected in customs and excise, plus any revenue surplus to its own modest, but growing, spending. In satisfying the fiscal needs of the newly federated states, the central government set its first tariff rates at something like the moderately protective Victorian rates. Radical 'liberals' like Alfred Deakin spuriously argued that the desire for tax revenue meant that protection must become the established policy of the Commonwealth, whereas much of the required tax revenue could have been obtained without the generation of significant protection against imported products, by attempting to balance excise taxes against tariff rates.

Table 2.2 Customs[1] revenues as ratios of imports, total taxation and GDP, Australia, 1850–1990

Year[2]	*Imports* (%)	*Total taxation* (%)	*GDP* (%)
1850	na	74	1.9
1860	na	90	3.9
1870	16	81	3.5
1880	17	74	2.9
1890	21	66	3.5
1900	19	64	4.1
1910	13	50	3.0
1920	10	23	2.4
1930	15	28	4.2
1940	16	22	3.5
1950	14	13	3.0
1960	9	5	1.2
1970	11	5	1.4
1980	10	5	1.3
1990	6	4	1.1

1 Customs and excise 1850–1900; customs only thereafter
2 Calendar years 1850–1900; fiscal years thereafter

Source: Derived from W. Vamplew (ed.), *Australian Historical Statistics*, Sydney: Fairfax, Syme & Weldon Assoc. (1987); Commonwealth Bureau of Census and Statistics, *Official Yearbook of the Commonwealth*, various; Reserve Bank of Australia, *Australian Economic Statistics 1949–50 to 1989–90*, Reserve Bank Occasional Paper No. 8 (n.d.).

The explanation of the protectionists' victory during the first decade after Federation mostly relies on an understanding of the various interests or aims that can be served by a tax on imports and the nature of the groups holding those interests. Here was a versatile policy instrument indeed, one that seemed capable of generating jobs, raising wages and rendering assistance to import-competing industries, even to labour-intensive export industries, while filling the government coffers. Not only were employers and employees in the protected industries possible winners from a policy of tariff protection but also were those on whom public revenue would be spent. The crucial alliance, however, was eventually formed between the parliamentary representatives of the labour movement and the protectionist forces, mostly from Victoria.

There was a temporary factor at work: protectionist sentiments were strengthened by the fact that the economy was weak, especially in Victoria, still recovering from the economic crisis of the 1890s, and severe drought in the eastern states had damaged primary as well as processing industries. Even at the Victorian tariff rates and excises, a common Australian tariff meant a reduction in the protection enjoyed by Victorian producers against producers in other states.

By themselves, the parliamentary representatives of the protectionist proprietors did not have a majority in parliament; they needed allies, some of whom they found in the labour movement. In the economic depression of the 1890s trade unions had suffered severe defeats, and so they looked to political power for help in achieving union recognition through the enactment of laws on compulsory arbitration. Under the new Australian constitution, the national parliament could make laws for 'conciliation and arbitration for the prevention and settlement of industrial disputes extending beyond the limits of any one State' (Australian Constitution, s.51 xxxv). Access was thereby won to another venue for the legislative support and nurture of unionisation, one that proved very important in later decades. In their search for parliamentary effectiveness, the Laborites made temporary coalitions with 'liberal' or progressive politicians. These latter frequently were protectionists.

In the first Federal Parliaments, the elected representatives of the labour movement mostly opposed the use of taxes on imports, fearing adverse effects on the cost of goods consumed by workers. Thus, the first federal tariff, that of 1901–02, could be considered only weakly protective. However, the Laborites soon became convinced that the tariff could be used for the improvement of workers' standards of living. The protectionist coalition of Laborite and liberal politicians defeated the free-traders in a decade that saw the emergence of the modern, disciplined, political party.

Possibly because competition was weak in the economy, it was widely believed that workers did not automatically share, in the form of higher wages, the advantage of the tariff protection that their employers enjoyed. Therefore in the mid 1900s, in a policy called 'New Protection', an attempt was made to link tax concessions to wages: 'The "old" Protection contented itself with making good wages possible. The "new"

Protection seeks to make them actual' (Alfred Deakin, quoted in Hancock 1930:83 from CPP 1907–08 vol. 2, pp. 1887–9). When they paid levels of wages approved by the Commonwealth Arbitration Court, individual employers in tariff-protected industries were to receive excise tax cuts. Because this form of discriminatory excise taxation was later declared unconstitutional, the wage-fixing tribunals had to be relied on as substitutes in the strategy to use tariff protection to assist unionised workers. At those tribunals, the payment of high wages became grounds for appeals for extra tariff protection; in turn, employers in industries enjoying higher rates of protection found themselves subject to claims for higher wages.

Among the losers from protection were citizens as consumers, a very diverse and ill-represented lot, as well as Australian exporting interests, especially pastoralists and miners, and those who relied on revenue from the land. Also harmed were employers and employees in firms, located mostly in Britain, that exported to Australia in competition with the protected Australian products. The Australian population was then mostly of British stock, much of it of relatively recent arrival. Many emigrant British workers received from Australian governments some subsidy towards the cost of migration, although the policy then was biased in favour of rural rather than urban immigrants. Given the continuing close ties that many Australians felt towards 'the old country', reinforced by the reliance that Australians had to Britain for defence (and *vice versa*), British interests were by no means completely ignored. The 1908 Lyne tariff, considered to be the first truly protectionist Australian tariff, provided for a measure of discrimination in favour of British exporters to Australia by way of preferential rates of tariff.

Of course, to the extent that the protectionist policy succeeded in creating more jobs in Australian cities for workers in the protected manufacturing industries, these jobs could be taken by British workers who, in effect, had been displaced by the Australian policy. Also, some foreign capitalists found it worthwhile, once the tariff was in place, to set up shop in Australia. Thus, by taxing the importation of products, the Australian tariff caused the 'importation' of workers and capital.

The triumph of the protectionists over the free-traders was not yet complete at the end of the first decade after Federation. Because the outlying states did not gain much by way of extra employment in the protected industries, they continued to complain about the policy. In particular, in Western Australia, primarily an agricultural and mining state, farmers rightly feared that the federal tariff would bring them the disadvantages of higher prices of farm machinery and higher wages of farm labourers. The state's voters had been attracted to the federal compact by a promise of a trans-Australian railway and by a constitutional provision granting Western Australia a period of five years outside the Australian tariff wall. In the 1930s the sectional disadvantages that the west suffered from protection became a cause there for efforts to secede from the Commonwealth.

Trade barriers were the most important but not the only means of industry assistance in the Australian strategy for economic development early in the century. For some industries, like iron and steel, a subsidy was a more attractive form of assistance; unlike a tariff duty, a subsidy did not harm the intermediate users (who were more potent politically than were final consumers). In addition, to the extent that the case for assistance to ferrous metals and metal-using industries rested on their potential importance in defence, a subsidy was a more efficient policy instrument than a tariff. For all activities competing with imports, the benefits of tariffs could be offset unless the rates of excise taxation on Australian production were kept below tariff rates.

The sugar industry received both a tariff and a subsidy, under circumstances that show the *direct* links that could be forged between trade policies and other policy objectives, in this case the White Australia policy. Sugar was grown in Queensland, using so-called Kanakas, who were mostly Melanesian labourers employed initially on indenture contracts. There was sufficient antagonism to black or coloured people in Australia at that time for a special deal to be struck with Queensland at Federation so that no new Melanesian labourers would be admitted and that efforts would be made to repatriate those already in the country. In return for this cost disadvantage, and presaging the strategy of 'New Protection', the Queensland industry in 1902 obtained a prohibitive tariff on the importation of sugar and remission of excise taxes for sugar grown by white labour (changed in 1903 to a bounty). Shann (1930:434–40) shows how governments became more involved: complete acquisition of raw sugar in 1915; government imports at a loss in 1919–20 and 1920–21; and exports subsidised by Australian as well as British consumers in the 1920s; but that is jumping ahead of the story.

'PROTECTION ALL ROUND' IN THE 1920S

After the interruption to international trade caused by the 1914–18 war, Australian trade policy became increasingly protectionist. Keith Hancock (1930:90–1) claimed that, in Australia, protection was more than a policy: it was a faith and a dogma; there was nothing surprising about the growth of protection, '[w]ith such a unity of spirit subsisting between private interests, public opinion, and the Commonwealth Treasury'.

The average tariff rate of the 1921 Greene Tariff was twice that of the Lyne Tariff of 1907–08. Furthermore, systematic extensions of protection were promised by the establishment of the Tariff Board. Most importantly, there was an extension of the protectionist umbrella to shelter the new chemical, engineering, transport and electrical industries. These industries and others had flourished during the hostilities when shipping was scarce and costly; imports hard to source; patriotic appeals made to consumers to support 'home' industry; and governments set up factories and gave orders to private factories for war-related commodities, including clothing. By these forms of temporary 'industry assistance', Australian producers of an import-competing product had been able to survive even while experiencing higher costs of production than incurred by the foreign rival. It proved hard for policymakers to allow the natural course of post-war economic events to reduce the 'war babies' in size. Those producing interests that thrived in temporarily favourable conditions of war incidentally obtained enhanced political or lobbying power, which they later parleyed into tariff protection in peacetime.

More generally, there is some evidence that part of the rise in the rates of the Australian tariff in the post-World War I period was a deliberate offset to the falls in natural protection. Indeed, the criteria developed by the Tariff Board in the 1920s, in advising the government on changes in the rates, involved comparisons between production costs in Australia and the CIF (cost, insurance, freight) or 'landed' cost of the imported (British) good. When substantial Australian production existed, even if stimulated by temporary conditions, the Tariff Board would be sympathetic to arguments that, through no fault of its own, the industry was hurting.

These were sectional policies, in so far as relatively few people in

Western Australia and Queensland, as well as farmers and graziers in all states, gained directly; indeed they lost from manufacturing tariffs, which raised prices of things they bought and created jobs elsewhere. However, not only manufacturing but also primary industries came increasingly to rely on tariff protection. This was the era of 'Protection All Round': tariff-supported arbitration for wage earners, tariff protection for manufacturers and tariff-supported marketing schemes for rural industry.

Until the 1930s there was strong political support in much of Australia, and less strong support in Britain, for schemes of Empire Settlement to place more British people on the lands of the dominions. In part, these schemes were a carryover of nineteenth-century strategies in eastern Australia of encouraging closer settlement on pastoral properties, reducing the political power of the squatter class, and a reflection of a prejudice against city life; in part, also, they were a desperate response to what was seen as a problem of employing soldiers, Australian as well as British, returning from war. To attract settlers, governments advanced loans to intending growers of wheat, sugar, fruits, meats and other labour-intensive products, lavished monies on infrastructure like railways and silos, and intervened in markets to enhance and stabilise incomes and prices. It was in these last efforts that the tariff and other trade policies played their central roles in the 1920s in supporting various agricultural marketing schemes.

For wool, wheat and other products of the land, the paraphernalia of rural statutory marketing schemes were developed during World War I: compulsory acquisition, growers' boards, fixed and pooled prices, advances before sale and so on. Using the powers of the Australian governments, the British contracted to buy huge quantities of Australian produce at guaranteed prices, some of them more favourable than post-war prices and some not; a vast stockpile of unsold wool was accumulated. During the 1920s varieties of marketing arrangements were developed, most based on use of the tariff to sequester the higher-priced home market from the foreign (usually British) market, paying the lower, world price. Additionally, the Australian Government pressed, somewhat successfully, for special access to British markets. In contrast to the United States' policy, Australian governments rarely restricted outputs and thus Australia helped drive down world prices.

As the Tariff Board set about its work of extending and administering protection, it eventually became concerned that the system had serious flaws. In effect, Australia had created administrative bodies to do the work of competitive markets (Capling & Galligan 1992:77–9); in the case of the Tariff Board, to safeguard the interests of consumers and exporters while at the same time giving a 'scientific' level of tariff protection to manufacturers and others. Particularly in 1925–27, the board complained of a tariff–wage spiral, alluded to earlier in this chapter: that with union support, it had recommended more protection to offset cost disadvantages in industries in which the wage-setting authorities subsequently granted wage increases that worsened the very cost disadvantage, so justifying another tariff increase; and so on. There were limits to its ability to spread the benefits of protection to the rural industries, caught in a squeeze of falling world prices and rising Australian costs. Typically, in response to these and other plaints, the Bruce Government commissioned a group of Australian economists, chaired by the protectionist J.B. Brigden, to inquire into the tariff. In the 1929 Brigden Report, the case was made that Australian protection supported the maintenance of a larger white population than could have been expected at the same standard of living (p. 140). Thus, the report went beyond endorsing the claims made by Syme and others, that protection could attract and retain population, to argue that Australian protection had in fact done so.

Without protection, average income per head would have been higher, it claimed, but the number of heads fewer and employed in the pastoral, mining and associated industries. However, the Brigden Report concluded that further extensions of the system of protection could not be justified; the strategy had reached its limit. (In arriving at this conclusion, the enquiry took informal account of the 'terms of trade' effect: that the Australian tariff, by reducing exports of wool, had increased its price, which was an economic benefit.)

The intellectual weaknesses of the Brigden argument are three-fold. First, it was based on a reading of Australian economic history that downplays the fact that, with little or no help from tariffs, in the last century Australia became a very urbanised, high-income society, one with an unusually high reliance on the tertiary or service sector for employment. There is a prejudice against services in the report, which biases its assessment of the alternatives to the tariff policy. Secondly, the report was unconvincing in its discussion of the redistributive mechanism through which import protection raised the level of workers' real wages: when could we be sure that the benefits from the *nominal wage* increases caused by the tariff outweighed the *consumer price* effects of the tariff? The so-called 'Australian case for protection' became the subject of international scholarly comment, and stimulated W.E. Stolper and P.A. Samuelson (1941) to publish one of the most highly cited economic journal articles in an attempt to remedy this defect in the report. Using a simplified model of comparative advantage, Stolper and Samuelson showed that tariff protection could raise real wages; on the same lines Mundell (1957) argued that if labour was thereby induced to migrate — the prime economic goal of Australian tariff protection — the tariff could, in view of the tariff-induced alteration in the pattern of comparative advantage, subsequently be removed and yet the additional population be retained. Thirdly, the report's argument, that the rise in the (real) incomes of workers was at the expense of a fall in average income, raises the question: Were there less costly ways to redistribute wealth and to attract population? An answer to that question had to await the tax-financed expansion of the social welfare system after World War II.

BOOST TO TARIFFS IN THE GREAT DEPRESSION

The cautious policy conclusion of the 1929 report was made irrelevant by the world economic crisis. Just as war produced a burst of protectionism, so did the cataclysmic disruption of the Great Depression of the 1930s. In the face of the worldwide depression, many nations practised 'beggar-thy-neighbour' trade policies, causing world trade to spiral downwards, dragging world production with it.

For Australians, the depression arrived in late 1928 and early 1929, with falls in world commodity prices and in the supply of foreign loans. Through the 1920s world agricultural and materials trade had been corrupted as many countries had artificially stimulated the output of primary products. By the end of the decade world commodity prices were falling. Adding pressure was the sharp fall in demand associated with the collapse of stock market prices in New York, and of banks in Europe and later in America. Australian governments, starting with the Federal

Government, soon encountered budget crises. They had borrowed very heavily during the 1920s for investments that would not have paid dividends even in good times. Now that both economic activity and prices were falling, their receipts from taxes or sales fell also, and fell sharply; their outlays were slower to adjust. They could no longer borrow the difference.

Hit first were Commonwealth customs receipts. In order to enhance federal tax collections, as well as to stimulate employment and to improve the balance of payments, emergency tariffs were imposed by the Scullin (Labor) Government in 1929 and repeatedly raised or extended. Various imports were prohibited, and a uniform tax, called a 'primage' duty, levied on almost all imports.

Such protectionist reactions were common and catching. Traditionally free-trade countries like Great Britain imposed new import protection. So concerned were more than a thousand American economists about the dire consequences of a savage increase in American tariffs that in 1930 they sent a public petition to President Hoover not to sign the Smoot–Hawley tariff into law; he signed. Nations retaliated against the policies of other nations, against devaluations and tariffs alike. Influenced by strong residual ties to Britain, and by the belief that recovery from the depression could be assisted by the formation of an imperial trade bloc (given structure by agreements signed in Ottawa, Canada, in 1932), governments in Australia and other British dominions discriminated against exports from the USA, Japan and other, third-world countries. The Ottawa Agreements were the subject of heavy criticism from the United States. Damage to international relations from spiralling trade restrictions and distortions, let alone the harm to the world economy, was so great that in 1934 the Roosevelt administration in the USA embarked on a crusade that was to have a huge influence after World War II, to build a world of freer trade and capital movements through the negotiation of bilateral and ultimately multinational treaties.

Undoubtedly, the number of urban workers on wages at the start of the depression had been boosted by the import protection accorded to manufacturing in the 1920s; in the slump, the unemployment rate among these workers was one of the highest in the economy. Whereas pastoral and other rural producers adjusted to a savage fall in prices by increasing the level of output, manufacturing employers cut output sharply and sacked workers. Employment in Australian manufacturing recovered early as the economy came out of the trough. Economic historians disagree about the role of manufacturing in the recovery and about the relative importance of the additional protection granted manufacturing by tariff increases and import restrictions, compared with the effects of currency depreciation and wage reductions (see Valentine 1987, Thomas 1988 and Spenceley 1990).

The emergency tariff measures of the Scullin Government were modified by succeeding governments. First, tariff rates applicable to British goods were reduced by a quarter, in keeping with the fall in the exchange rate of 1931. Next, the Tariff Board assumed a central role in the administration of the Ottawa Agreements, which was to recommend tariff rates that compensated those Australian industries it judged to be 'reasonably assured of sound opportunities for success' for their higher costs of 'economical and efficient production'; as well, the board recommended the tariff preference in favour of UK producers. Although the board generally proposed rates lower than the Scullin rates, the British governments were not fully satisfied with Australian compliance with the Ottawa Agreement. In particular, the Australian Government refused to treat all British preferential rates as 'bound' (i.e., as maxima), asserting that if the wage-fixing authorities increased Australian costs, then the board might have to recommend a tariff increase (Drummond 1974:408). Rates were also increased in 1936, without a prior Tariff Board reference, during the damaging episode of trade diversion against Japanese textiles and American cars. The net effect was an average Australian tariff higher in 1939 than before the depression.

What Australians, along with producers in the other dominions, received in return for the tariff concessions of their government at Ottawa was preferential access to UK markets for a range of primary products. This access was most valuable to Australian exporters of meat and (to a lesser extent) of dairy and fruit products. Although the British Government refused what in effect were requests to guarantee the Australian living standard through bulk purchases at fixed prices (Drummond 1974:158), in the 1930s it forced UK consumers to subsidise the products of the land settlement schemes that the British Government had subsidised by loan and grant in the 1920s. Here indeed was protection all round, or at least protection of the labour-intensive sectors: persuading the British not only to accept tariffs on British manufacturing exports to Australia, in order to assist Australian manufacturers, but also to grant preferential access to Australian labour-intensive primary exports (see Tsokhas 1992; Ross 1990 argues that the tariff strategy was substantially motivated by defence considerations).

An example of the use of trade restrictions for ulterior motives is the ban on the export of iron ore Australia imposed 1938. Huge but (at that stage) uncommercial ore deposits in north-west Western Australia, at Koolan Island in Yampi Sound, had been taken over by a British firm, Brasserts, fronting for a Japanese steel producer with close ties to the Japanese Government. Alerted to the strategic disadvantage of having Japanese engineers and nationals in the area, the Australian Government searched for an excuse to remove them. It seized on the notion that iron ore was scarce, at least in commercial quantities and qualities, close to shipping. The head of BHP, Essington Lewis, did not believe it; nor did the Japanese, who protested that, even if the claim was correct, for the sake of a cordial international order countries like Australia with huge natural resources should not deny them to the 'have not' nations. (The ban on ore exports lasted until the early 1960s and delayed the rediscovery and exploitation of huge West Australian ore deposits.)

POST-WAR RECONSTRUCTION: IMPORT REPLACEMENT

In the period of reconstruction after World War II, the slogan 'Populate or Perish' gained fresh voice in Australia. With the displaced persons of Europe as a new source of immigrants, Australian trade protection was broadened and escalated to give them jobs. An urgent policy of development was attempted against a background of balance of payments stringency induced by fixed exchange rates. Therefore, in the 1950s quantitative import restrictions (or quotas) temporarily surplanted tariffs. Coalition Governments involving the Country (now National) Party won over agricultural but not mining interests by various stratagems, including the dismantling of the preferential trade relationship with the United Kingdom and new trade treaties with the country fast becoming Australia's biggest export market, Japan.

Interestingly, although Australian politicians and bureaucrats had pranced on the world stage during the 1940s, advocating international collective strategies for full employment, Australian trade policy in the 1940s, 1950s and (decreasingly so) in the 1960s was held largely aloof from the American-led thrust towards a more liberal world trade regime. Australian politicians and bureaucrats pleaded that Australia was a developing country and, as such, in the international fashion of the times, wished to pursue a strategy of import replacement. Australian governmental representatives also argued that its various schemes to support the 'home' prices of agricultural products Australia exported should not be subject to being negotiated away under rules of the General Agreement on Tariffs and Trade (GATT), which was signed in 1947. Also, in the first decade of GATT, Australian primary exports benefited somewhat from a series of bulk purchasing agreements with the UK Government, mostly begun in World War II (Crawford 1968, section 7). The engine of world economic growth was trade, but the Australian economy, although it benefited from the trade-induced spurt in the world economy, was partly decoupled from that motor in the 1950s and 1960.

None the less, the Australian economy grew and prospered in the 'long boom' of the late 1940s to the early 1970s; it might have grown more and prospered more if trade and financial flows had been freer. The share of manufacturing industry, the chief beneficiary of protection, in national output and employment reached historic peaks in the late 1960s.

AN END TO TARIFF PROTECTION?

Still, the belief that good jobs could be created through a combination of import protection and wage regulation began to fade, for the evidence was that the more highly trade-protected was an industry, the lower its human capital and wages. Decades of a 'made-to-measure' tariff system had produced Australian entrepreneurs more skilled in 'rent-seeking' in Canberra than in generating good jobs by 'profit-seeking' in international markets. The costs of an inward-looking economic strategy were increasingly obvious, especially after the theoretical work of W.M. Corden, the parliamentary speeches and journalistic efforts of C.R. 'Bert' Kelly and the bureaucratic heroism of G.A. Rattigan.

In the 1960s the tariff had again become the chief means of affording import protection. The board took the approach that it had used in the pre-war period: recommending the rates needed to offset the cost disadvantage of the Australian industry. A political alliance was formed between the Country Party and the protected, mostly urban interests (Glezer 1982). When the Vernon Committee reported in 1965, it criticised the wide dispersion of rates of protection, especially when judged by the measure that Corden had advocated, the effective rate of protection (see the introduction for definition); it suggested that a benchmark rate be used of, say, thirty per cent. Instead, the Tariff Board, under the chairmanship of Rattigan, embarked on a strategy designed to bring down the high rates, with the first step being a review of the tariff, which actually did not begin until 1971; previously, the board had only conducted inquiries on 'references' from the government minister. In the meantime Rattigan set about increasing the economic expertise available to the board, both staff and consultants, and arranged for the collection of much more information than the board previously had at its disposal. In the coming battle over protection the board would be well armed with data, models and analysts. From the board's investigations important facts or arguments emerged. First, in the 1960s the pattern of protection was still heavily influenced by the 'emergency' tariffs of the depression. Secondly, most manufacturing employment was in industries or activities enjoying relatively low rates of effective protection. Many of the highest effective rates depended on the use of 'by-law' admission of duty-free parts for which there were no close Australian substitutes and on associated 'domestic content' plans (Rattigan 1986). In its annual reports, the board argued that none of the standard economic arguments used in favour of protection provided sufficient justification for these large variations in rates of protection. Economic analysis indicated that there were large efficiency gains to be achieved if the high tariff rates were reduced.

Boosted by the presence of experts in labour economics, such as Robert Gregory, the board began to provide answers to the question: Where would the alternative jobs come from? Drawing on reports and research papers that stressed the high rates of labour turnover in the Australian job market, and the costly fragmentation that trade protection had caused (in 'white goods' and motor vehicle production, in particular), the board and commission claimed that tariff reductions were a minor cause of job loss. More fundamentally, Rattigan criticised the question as revealing a confusion between techniques of detailed economic planning — which was what the board did in earlier decades — and what was later to be called 'levelling the economic playing field'. The board now would recommend lower rates of protection, unless there were convincing and publicly argued justifications for special treatment, and these cases had to be more than the incantation of words like *market failure* or *externalities*. Despite these attitudes, Rattigan realised that the board and commission could blunt attacks by providing the results of general equilibrium and other investigations of the effects of tariff cuts on employment, output and so on; thus, he launched the heavily funded 'Impact' Project.

Rattigan's ideas found sympathetic ears in the Whitlam Labor Government elected in 1972, for its leaders believed that the system of protection had helped their political opponents and became convinced that a reduction in protection would help to generate the larger economic 'cake' from which an ambitious program of social spending could be funded without excessive rises in general tax rates. The Tariff Board transmogrified in 1973 into the Industries Assistance Commission, still under Rattigan, with the purview extending to all sectors. A major turning point was the 25 per cent tariff cut in 1974, adventitiously undertaken by a Labor government to increase commodity supplies in a boom so as to dampen inflationary pressures. Although the cut was blamed by some for the collapse in economic activity in 1974–75, although the system of paying 'adjustment assistance' to displaced workers and firms proved less satisfactory than expected, and although some industries hardest hit by the across-the-board tariff cut subsequently obtained less transparent forms of assistance, the protectionist tide had turned.

In three major reports in the second half of the 1970s, which contributed to the flow against protection, we can find elements adapted by the Labor governments of the 1980s to smooth the transition to lower rates of protection or industry assistance: from the 1975 Green Paper, the proposals about participatory decision-making process, which became the industry plans of the 1980s; from the 1977 White Paper, the necessity of a definite commitment to tariff reductions as the long-term goal; and from the 1979 Study Group, an emphasis on export incentives. Industry plans, for autos, steel, heavy engineering, and for footwear, clothing and textiles, offered these highly protected industries some temporary exemption from tariff reductions and provided them with some other forms of assistance in order to bring about sufficient structural adjustment to render them less reliant on protection. Along with other instruments of 'adjustment assistance', the industry plans could be seen as a tactic of compensating the heaviest losers and rendering them politically less dangerous to the strategy of lowering Australian trade barriers.

The Hawke Government, backed by its Accord with the unions, embarked on its program of tariff reductions in favourable economic circumstances, after the substantial fall in the value of the Australian dollar and in a boom; its successor, the Keating Government, convinced that an open economy was more productive and adaptable than a protected one, accelerated the program in the deepest recession since the 1930s (Grattan & Gruen 1993; Capling & Galligan 1992, especially chapter 4). Thus abandoned was the defensive Australian historic compromise of social protection against the market (analysed in Castles 1988), except in the one market at its very core, the labour market. Severed was the link that had been forged in the nineteenth century between a protective trade policy and job creation.

FURTHER READING

Introductions to the historical context are to be found in Hancock (1930), in Butlin, Barnard and Pincus (1982, sections by Butlin) and in Shann (1963). All emphasise the interrelationship between trade and other policies, especially immigration and wage regulation; see also Castles (1988). Patterson (1968) surveys writings on the later nineteenth century, while Reitsma (1960) also covers the first half of the twentieth. An excellent, more recent discussion is provided by Anderson and Garnaut (1987), while Smith (forthcoming) breathes life into the topic. Rattigan (1986) gives an insider's view of the start of the process of tariff reduction, of which process Bell (1993) does not fully approve.

3

Policies affecting Australia's primary sectors

Kym Anderson and

Christopher Findlay

Significant changes have occurred over time in the share and composition of primary products in Australia's exports. Until the early 1960s more than 90 per cent of exports from Australia were primary products, mostly from farms (except during the gold rush periods). Now primary products account for about two-thirds of Australian exports, and farm products are only half as important as exports from mines. Even so, these are still very high shares compared with other industrial countries. In fact primary products contribute less than a third of global trade, or less than half their share in Australia. That is, Australia has retained a strong comparative advantage in primary products.

Why is Australia's comparative advantage still in primary products? What has determined the pattern of trade specialisation *within* the primary product sector, including the extent of processing before export? In particular, what have been the contributors to and the impacts of the mining booms since the 1960s? How have changes in trade and other producer support policies at home and abroad affected primary exports? These are key questions addressed in this chapter. There is also some discussion of taxation arrangements affecting the minerals and energy sector, of rules on access to land for exploration and mining, and of the interrelationships between environmental and trade policies which affect both agricultural and mining industries. The chapter concludes by drawing out the implications of continual structural and policy changes for the future of Australia's primary product trade.

AUSTRALIA'S EXPORT SPECIALISATION: WHY IS IT STILL IN PRIMARY PRODUCTS?

It was noted in chapter 1 that historically Australia's exports have been overwhelmingly primary products. Their share has fallen since the 1950s, but in 1992–93 they still accounted for nearly two-thirds of the value of exports (including some semi-processed non-ferrous metal products) and well above the value of imports of primary products (see tables 1.1 to 1.3).

The sectoral composition of primary product exports, as reported in table 1.1 and illustrated in figure 3.1, also has changed. The rural product share of exports has fallen from 86 per cent to 23 per cent in the period since 1950 (much more than the fall for the rest of the world), while the minerals and energy sector's share has risen from 6 to more than 40 per cent (or from well below to well above the share of these products in global trade).

The main explanation provided in chapter 1 for this concentration of exports in resource-based (or primary) products is the relative abundance of natural resources per worker in Australia. This point was demonstrated by comparing Australia's factor endowment ratios with those of other economies, as depicted in the Leamer triangle in figure 1.1. But relative factor endowments provide only part of the explanation for the pattern of trade specialisation. Another important part is the pattern of distortionary price and trade policies at home and abroad. In the absence of its protectionist policies towards manufacturing, Australia might have had an even stronger export specialisation in primary products historically (Anderson 1987a; Anderson & Garnaut 1987). And its exports certainly would have remained more specialised in farm products and coal had other OECD countries not increasingly assisted their farm and coal-mining industries in recent decades (Tyers & Anderson 1992; Jolly, Beck & Savage 1990).

It is true that some agricultural industries in Australia have enjoyed considerable assistance from governments, and on average the farm

Figure 3.1 Composition of Australia's exports, 1950–51 to 1992–93 (%)

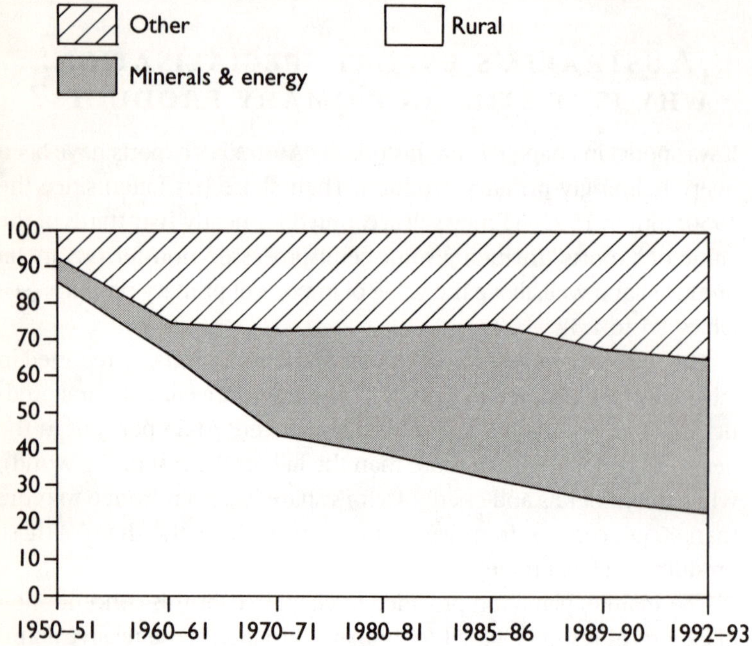

Source: Australian Bureau of Agricultural and Resource Economics, *Commodity Statistical Bulletin 1990* and *Agricultural and Resources Quarterly*, December 1993, Canberra: AGPS.

sector now receives no less effective assistance than manufacturing, following the decline in Australia's tariffs and the abolition of its import quotas on manufactures (table 3.1). Hence if Australia were to adopt free trade in the years to come, agricultural exports might not have a much greater share of total exports than they have now.[1] The share will be somewhat higher, though, thanks to the success of the GATT Uruguay Round of multilateral trade negotiations (1986–94) in so far as that results in a reversal of the long-run growth in agricultural protectionism in other OECD countries.

Table 3.1 Effective rates of assistance for Australian agricultural industries and for mining and manufacturing, 1970–91 (%)

	Dairy	*Eggs*	*Tobacco*	*Dried vine fruit*	*Extensive grazing*	*Total agric.*	*Total mining*	*Total manuf.*
1970–74	65	>100	>100	43	6	22	na	33
1974–78	73	>100	>100	31	7	11	na	26
1978–82	70	>170	>220	3	7	9	na	23
1982–86	130	>125	91	62	6	13	–4[1]	21
1986–90	125	34	72	49	6	12	–3	17
1990–91	73	34	36	70	14	15	–3	14

1 1983–84 only

Source: Industry Commission, *Annual Report 1992–93*, Canberra: AGPS (and earlier issues).

Meanwhile, effective assistance to mining has been negative historically. It averaged around –3 per cent of value added in the 1980s and early 1990s (Industry Commission 1992:295), but was much more negative in earlier decades, particularly for industries subjected to export prohibitions such as iron ore in the 1940s and 1950s. Hence the recent growth in importance of mineral and energy exports is in part attributable to the gradual but sustained changes in Australia's trade policy regime (the declines in both the negative effective assistance to mining and the positive assistance to manufacturing). The residual part of the explanation has to do with changes in relative prices of Australia's exports in international markets and the not-unrelated discoveries and exploitation of new mineral deposits, discussed below.

There are important differences between the primary sectors' shares of exports and their shares of output and employment. For example, the farm sector in 1992–93 accounted for 23 per cent of exports but only 5 per cent of employment and 3 per cent of national output. The mineral resources sector accounted for more than 40 per cent of exports but only 2 per cent of employment and less than 7 per cent of national output. That is, the part of the economy that accounts for nearly two-thirds of Australia's exports directly provides only about one in twenty jobs; the rest of the labour force is employed in manufacturing (around 15 per cent) and, most importantly, in the service sector (the remaining 80 per cent). Moreover, these statistics imply that primary production is much less labour-intensive than non-primary production, as its 10 per cent of GDP is produced with only 7 per cent of the workforce.

The relative labour intensity of non-primary activities, combined with the economies of scale and agglomeration in their production, explain an important apparent contradiction in Australian society: its very high degree of urbanisation, notwithstanding its strong comparative advantage in resource-based products.[2] The important connections between the (largely non-urban) resources sector and the (largely urban) non-traded goods and services sectors are discussed below, before turning to examine the determinants of export specialisation *within* the primary product group.

WHY IS RESOURCE-RICH AUSTRALIA SO URBANISED?

Australia always has had very few people relative to natural resources. That shortage of labour, together with abundant access to financial capital (first from Britain and more recently, from the United States and Japan), ensured high real wages and hence the creation and/or adoption of the most capital-intensive (labour-saving) forms of production in each industry. Being perhaps the highest-wage country in the world until early in the twentieth century (see chapter 1), it was in Australia's economic interests to develop new labour-saving technologies. In agriculture that manifested itself in, among other innovations, the designing and manufacturing of larger farm machinery capable of coping with the relatively rough terrain, and of complementary innovations to crop and livestock practices (Donald 1982). And in mining, too, highly capital-intensive methods were and continue to be adopted, especially in the more remote locations. Hence relatively few workers are needed on Australia's broadacre farms[3] and mine sites.

But that is only part of the reason for Australia having most of its population in the urban coastal fringe. A more important part of the explanation is that many of the purchased inputs for those primary activities, and the services required to get their output to market, can be provided most efficiently from Australia's large urban centres adjoining ports.

Furthermore, as technological changes occur in the primary sectors, or there is a new minerals discovery, or the export price for part of the sectors' output takes a seemingly permanent jump (as occurred with energy raw materials in the 1970s, thanks to OPEC), the sectors producing non-tradables — which include many urban service sectors — typically receive a boost. The reason for that can be seen with the help of figures 3.2 and 3.3.[4]

Consider the case of a new minerals discovery (or the discovery of a new technology such as a cheaper extraction method). The announce-

Figure 3.2 Impact on the national market for non-tradables of a new minerals discovery

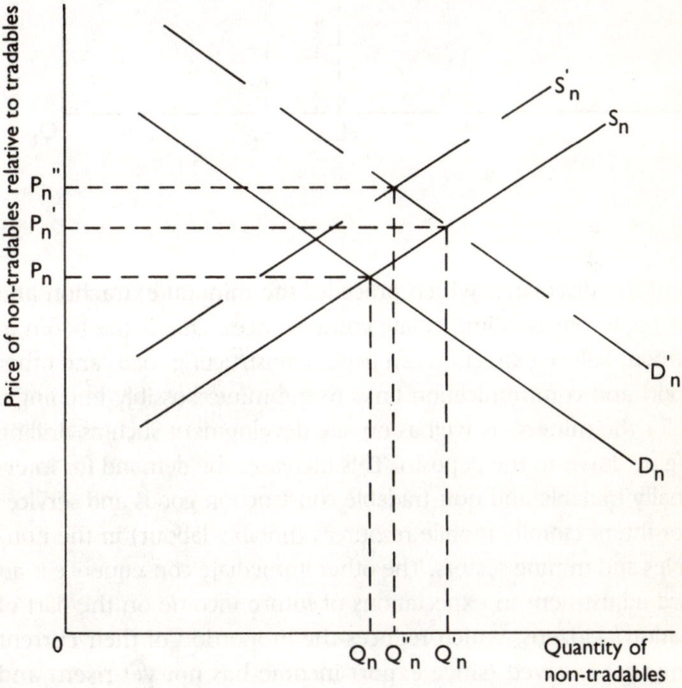

D_n and S_n are the demand and supply curves for non-tradables.

Figure 3.3 Impact on the national market for labour of a new minerals discovery

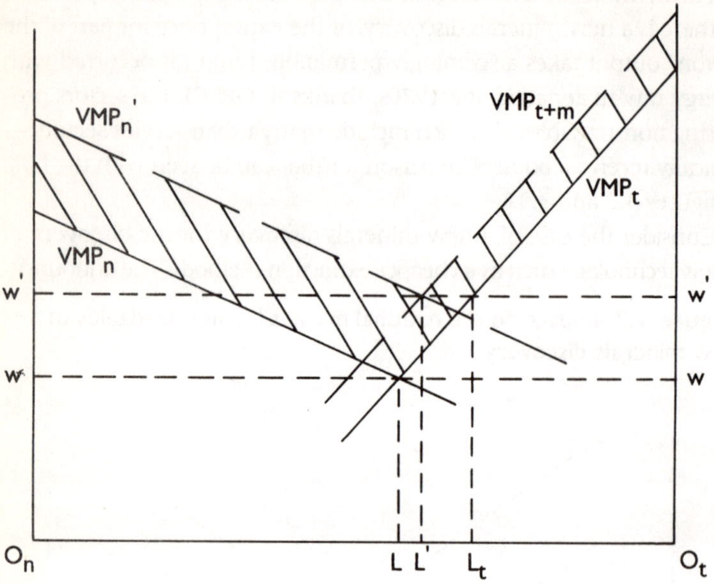

ment of the discovery, which precedes the mineral extraction and export boom has two immediate consequences. One is the boom in investment before extraction can begin (constructing roads and other transport and communication links to the mine, possibly building a town for the miners, as well as on-site development such as drilling or digging down to the deposit). This increases the demand for internationally tradable and non-tradable construction goods and services and for intersectorally mobile resources (notably labour) in the non-tradables and mining sectors. The other immediate consequence is an upward adjustment in expectations of future income on the part of the nation's citizens, which reduces the proportion of their current income that is saved (since export income has not yet risen) and increases their demand for tradable and non-tradable consumer goods and services.

Both spending effects thus increase the demand for non-tradables (and for foreign savings) and thereby cause the price of non-tradables relative to tradables to rise, *ceteris paribus*. In figure 3.2 the demand curve for non-tradables shifts from D_n to D_n', forcing up the relative price of non-tradables from P_n to P_n'.

The resource movement and spending effects, involving among other things the movement of labour away from the production of other tradables (the price of which is given in international markets and assumed not to change) and into mining development and the production of non-tradables (the expanded demand for which can only be met by domestically produced supplies), are also depicted in figure 3.3. Before the discovery the demand curve for labour (the marginal product of labour times the price of the sector's output) in the non-tradables sector is represented as VMP_n, while the mirror image of the demand curve for labour in the tradables sectors is shown in figure 3.3 as VMP_t, where the length of the line O_nO_t represents the aggregate supply of homogeneous labour (assumed to be constant). Labour in the non-tradables sector is measured to the right of origin O_n and labour in the tradable sector is measured to the left of O_t. Assuming no distortions in the labour market and free mobility of labour intersectorally, the nation's labour force is fully employed at a common wage, w in figure 3.3, and O_nL/O_nO_t is the proportion of the labour force employed in the non-tradables sector. The increase from P_n to P_n' due to the spending effect on the demand for non-tradables boosts the demand for labour in the non-tradables sector. And on the tradables side of figure 3.3, a demand for more labour in the mining sector shifts the total demand for labour in tradables sectors to VMP_{t+m}.

As a consequence of these two shifts the real wage rises, raising the marginal cost of producing non-tradables (the shift from S_n to S_n' in figure 3.2). Hence their price rises, to P_n''. The net effect of these changes is to shift the demand for labour in non-tradables to VMP_n' and the wage to w' in figure 3.3. Employment in the traditional tradables sectors (assuming that the already existing mining projects bear none of the adjustment) shrinks to $L'O_t$, the mining sector employs an extra $L'L_t$ workers, and employment in the non-tradables sector changes to O_nL', which might be more or less than O_nL in figure 3.3 (just as the quantity of non-tradables now produced, Q_n''', might be more or less than the pre-discovery quantity of Q_n in figure 3.2). Real domestic product has increased, by the shaded area between the Vs formed by the solid and broken labour demand curves of figure 3.3.

In summary, real income has increased but resources also have been reallocated towards the minerals component of the tradable goods sectors and away from the traditional producers of tradables. The effect on employment in non-tradable sectors could be positive and negative, but is more likely to be positive the smaller the increased demand for labour in the mining sector (e.g., because of the use of ever more capital-intensive techniques).

Figures 3.2 and 3.3 can be used similarly to analyse the effects of a change in the price of tradables internationally (or domestically because of a change in trade policy), for that too shifts out both the demand for non-tradables in figure 3.2 (assuming non-tradables and tradables are substitutes) and the demand for labour in tradables production in figure 3.3. Given the combination of Australia's resource discoveries, its trade and industry policy changes in favour of resource-based industries and the improvement in international prices for energy raw materials in the 1970s, the changes in the mix of Australian exports and in the composition of GDP and employment in Australia (see tables 1.1 and 1.7) following the series of big mining booms in the late 1960s and beyond are understandable.[5]

MORE ON THE DETERMINANTS OF THE COMPOSITION OF PRIMARY PRODUCT EXPORTS

As already indicated, a number of variables affect the composition of the natural resource-based group of exports. First, as was pointed out in chapter 1, there has been a long-term decline in the prices received for farm products on world markets (see figure 1.3). This decline is due in part to differences in shifts in the demand for and supply of agricultural products; despite the growth in world population and incomes, the demand for food has grown more slowly than supply, reflecting rapid technological changes on the one hand and the relatively low income elasticity of demand for food products on the other (Anderson 1987b).

A further contributor to the long-term decline in farm product prices is the growth in agricultural protection in the rest of the world. The

Table 3.2 Nominal rates of assistance to agriculture in industrial countries, 1979–93 (percentage, based on producer subsidy equivalents)[1]

	1979–81	*1982–84*	*1985–87*	*1988–90*	*1991–93*
EC-12	58[2]	50[2]	85	78	90
EFTA	na	113[3]	180	230	246
Japan	133	170	250	184	187
United States	19	28	56	27	26
Canada	32	41	79	54	53
Australia	10	16	16	9	11
New Zealand	22	37	28	6	6
ALL OECD	**41**	**45**	**79**	**64**	**69**

1 The per unit producer subsidy equivalent, expressed as a percentage of the border price for each commodity and averaged across commodities using production valued at border prices as weights. The EC rates include national government assistance as well as that provided by the community's CAP.
2 EC-10.
3 1979–85.

Source: OECD (1994).

gap between domestic prices in protectionist countries and international prices has increased throughout the post-war period but especially since the early 1980s (table 3.2). Surplus production generated in protectionist countries has been sold on world markets with the help of export subsidies. Australia has suffered doubly from this: from denial of access to markets in protectionist countries and from lower (and more volatile) prices in the residual international market (Tyers & Anderson 1992).

There are political economy reasons for expecting a systematic tendency for governments to protect farmers increasingly in the course of economic development, at least up to some threshold level (Anderson 1994a). Agricultural protection in advanced industrial countries might therefore be considered normal rather than exceptional. Liberalising those policies would be unattractive politically — unless new influences, such as pressure from abroad, upset the domestic political equilibrium. There have been recent examples of such influences operating in bilateral relations, such as United States pressure on food policies in Japan and Korea, and the Uruguay Round of GATT multilateral trade negotiations. Other changes underway in Europe are also lowering the political cost of reforming Western European policies (Anderson 1994b). Even so, reforms to farm policies abroad during the years ahead are likely to be too slow to prevent the relative price of farm products in international markets from continuing to trend downwards. Without increases in government assistance to agriculture in Australia, the relative contributions of the farm sector to GDP, employment and even perhaps exports will therefore probably continue to decline over time, particularly if the remaining reductions in protection to our manufacturing industries remain phased over a long period.

One industry — wool — responded to a drop in its export price in the late 1980s by choosing to limit its exports by borrowing to purchase offerings that did not fetch above a certain floor price at auction. The resulting purchases were stockpiled, in the hope that this would raise the international price enough for those stocks eventually to be resold at a price that would more than cover storage and interest charges. That action turned out to be about as effective as trying to prevent the tide from coming in: at the time of writing, several years later, much of the stockpile remains unsold, and the price of wool remains well below the peak of 1987–88 that growers had vainly aspired to maintain.

International price changes, resource discoveries and changes in domestic policy have all influenced Australia's supply of minerals and energy exports in important ways. One obvious price change has already been mentioned, namely, the rise in the relative price of energy raw materials during the 1970s as a result of decisions by OPEC to reduce output. The price of coal in particular would have risen even more had it not been for the rise in subsidies and import protection for coal-mining in Western Europe and elsewhere (Jolly, Beck & Savage 1990; Steenblik & Wigley 1990). As with agricultural protection, though, there are signs that coal subsidies might plateau or even begin to be reduced soon, again partly as a consequence of commercial diplomatic pressure from major exporting countries such as Australia.

Another important influence on the mining sector's export prices has been the dynamic growth in resource-poor East Asia (first in Japan and later in other East Asian economies) in their demands for mineral and energy raw materials. Australia, with its locational advantage relative to other, more distant suppliers, has been able to sell into those markets at prices above those previously received net of transport costs. These developments stimulated much greater exploration efforts to find new deposits for export markets (Doran 1984; Freebairn 1987). The volume of exports of minerals increased dramatically as a consequence. Even by the mid 1970s coal had replaced wool as Australia's single largest export commodity in value terms.

Demand expansion abroad was not the only influence on the mining sector, however; Australia's own mining policies also played a significant role. Smith (1983) points out that before the 1960s a policy goal in Australia was to ensure the availability of 'key' resources to supply the growing domestic demand. In some cases, this led to tax concessions and subsidies to stimulate exploration (e.g., for oil). In other cases it led to a ban on exports (e.g., of iron ore between 1938 and 1960 and manganese for a shorter period). Needless to say, the export embargo deterred exploration. So too does the policy that bans the opening of any new uranium mines — a policy that has been in place since the early 1980s in response to pressure from the anti-nuclear lobby.

Australia's policy regime has been such that, as already noted, the effective rates of government assistance have differed markedly among tradable goods sectors. This has induced transfers of productive factors between them. The rates of assistance for agriculture and manufacturing have both been positive, the latter traditionally much larger, but it fell over the 1980s so that it is now the same as that for agriculture (see figure 1.6a and table 3.1). The effective rate of assistance for the mining sector, in contrast, has always been negative and substantially so before the 1960s. The policies responsible for that inter-sectoral pattern of assistance have therefore discouraged the minerals sector and especially mineral exports in the past. However, the steady falls in rates of protection for manufacturing in the 1970s and 1980s, and further falls expected over the 1990s, have and will continue to reduce the extent of this distortion against mining.

A commonly asked question is: why does Australia export such a small share of its minerals and energy in processed forms? As figure 3.4 shows, metals have contributed less than a twelfth of the country's merchandise exports during the past fifteen years, while ores and energy raw materials have each been as much as twice as important.

It happens that several factors affect the location of minerals processing. One has to do with comparative advantage. Contrary to popular belief, it need *not* be the case that a country like Australia,

Figure 3.4 Shares of ores, fuels, gold and metals in Australia's total exports, 1978–79 to 1992–93 (%)

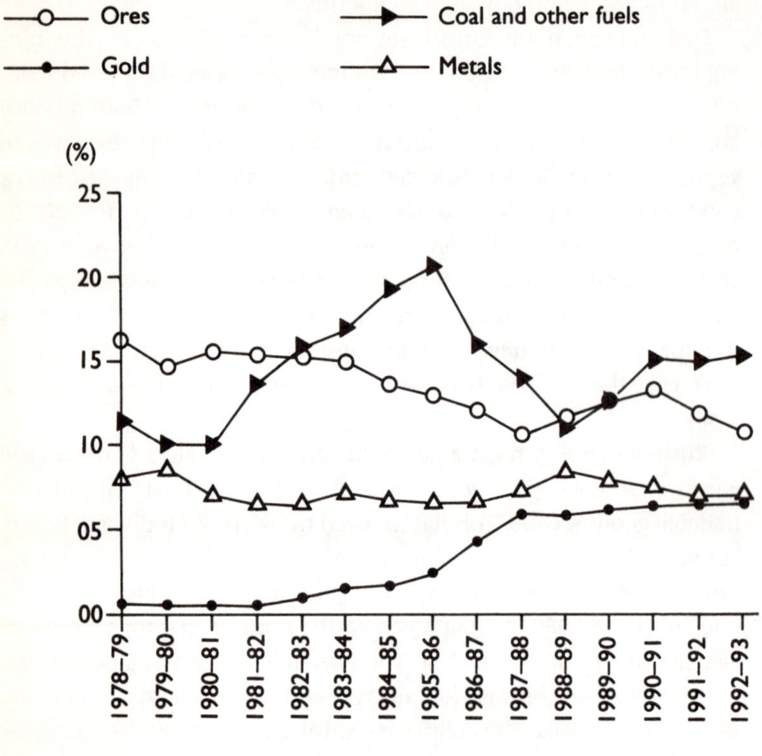

Source: dX Software database (EconData Pty Ltd, Armadale Vic)

which is well endowed with raw materials, is the most efficient location for processing those materials. The saving in transport cost from exporting the processed rather than unprocessed product might well be insufficient to offset the higher cost of processing here compared with processing in lower-wage countries.

As with agriculture, another factor reducing Australia's metal exports has to do with government policies abroad. Glance, Johnston and Winton (1992) argue that in East Asia border protection measures are particularly important in the minerals (rather than energy) trades. Furthermore, the importance of these barriers tends to rise with the degree of processing. This feature tends to bias the import composition of those resource-poor economies towards raw materials rather than processed products. Reducing these barriers to trade with East Asia would provide important benefits for Australia, such as in larger exports of steel products.

In addition to these demand-side factors, the extent of processed minerals product exports also depends on supply-side conditions in Australia. The latter include not only natural comparative advantage factors but also other aspects of costs, most notably Australia's regulatory and tax policies. While there has been a package of policies on the agenda for so-called microeconomic reform for more than a decade, the pace of such reform has been relatively slow in many areas (see, for example, Forsyth 1992). A speeding up of those reforms would undoubtedly help Australia to become more internationally competitive in many industries, not least in metals manufacturing and other processing of primary products.

OTHER PRIMARY SECTOR POLICIES AND THEIR IMPACT ON TRADE

Taxation arrangements and environmental policies have been important issues in the Australian primary product sector, especially in minerals and energy (Smith 1983) but, in the case of environmental policies, increasingly also in agriculture (Chisholm 1992). How have they impacted on Australia's primary product trade?

The motivation for special taxation arrangements for the minerals sector was that mineral deposits were regarded as a valuable asset that belonged to the whole community. The community, it is argued, should be compensated for the depletion of and the reduction in the value of this asset that occurs as a result of mining. Without such a mechanism in place, shareholders of the mining company would gain free access to this asset.

Smith (1983) reviews the various mechanisms by which the community might gain some compensation and their distorting effects. A common form of taxation has been a royalty (set, for example, in terms of dollars per tonne of mine output). The royalty creates a distortion because it affects decisions on mining at the margin. Annual output of a mine is reduced, and mines are closed earlier than otherwise. Thus under this tax regime the mining sector is smaller than otherwise. Less distortive would be a tax on profits in excess of a specified rate of return (one that reflects appropriately the risks involved in the mining industry) for this would not affect marginal decisions on mining practices.

The taxation regime is of course not the only force that affects the rate of resource depletion. An excessive degree (from society's point of view) of risk aversion also would lead to less investment in risky activities. However, the effects of risk aversion on output are not clear. On the one hand, there could be too little exploration activity, and some deposits would remain undiscovered because exploration is risky; but, on the other hand, holding a deposit for later use also is a risky activity, so existing deposits might be mined *more* rapidly than if society was less risk averse.

Some institutional features of the way that mining leases are administered by governments can have the same effect. An example is a set of rules requiring that a deposit once discovered must be exploited within a particular period or else relinquished. The optimum strategy in terms of wealth maximisation (that is, solving the profit maximisation problem over a long time period) might be to wait while higher-quality deposits elsewhere are exploited first and prices rise as those deposits are exhausted.

Access to mineral deposits also depends on the definition and administration of the property rights of the traditional owners of the land involved. Those rights were redefined in the Mabo legislation passed in late 1993. Existing mineral leases are not affected, but new projects or renewals of leases will involve negotiation with native title holders. Lawyers have argued that this legislation increases the uncertainty associated with future minerals projects in Australia. It is also argued that the legislation creates much more uncertainty for the minerals sector (and more than it creates for the pastoral industries where all existing leases will extinguish native title and where no new areas of land are expected to be opened up).[6] If this is the case, and if the outcome depends on the results of test cases or challenges to the legislation by state governments, then the legislation has an effect akin to a loss of comparative advantage in mineral activities. Australian mining firms will in that case shift a greater share of their exploration and mining investment activities to other countries.

The higher profile of environmental issues in policy debates is having both positive and negative effects on output. Certainly as environmental policies imposed in Australia raise costs of production, they will weaken the ability of our industries to compete internationally. But insofar as similar policies are being imposed abroad, competitors' costs are being raised as well. The net effect of those parallel developments might well be an improvement in Australia's overall competitiveness and in particular that of its primary sectors. This is because Australia is relatively well endowed with environmental resources, especially compared with its more densely populated trading partners in East Asia. Hence, as environmental resources in those economies are diminished as a result of economic growth, and as local communities' evaluation of the natural environment increases, Australia's comparative advantage in activities that use the natural environment intensively, such as farming, mining or minerals processing, will increase (Anderson 1993).

A caveat is in order, though. With the growing interest in environmental protection policies comes a danger that environmental groups will perceive trade policies as a potential stick or carrot for imposing their views about 'appropriate' environmental policies on other countries. This is of concern because, if unchecked, it could lead to retaliation, counter-retaliation and ultimately an undermining of the GATT rules-based global trading system on which Australia's prosperity depends (Anderson & Blackhurst 1992).

CONCLUSION

Australia's export specialisation in primary products is due mainly to the relatively large endowment of resources per worker in Australia. The substantial changes in the mix of primary product exports since the 1960s, in particular the decline in the rural share and the rise in the minerals and energy share, have been associated with changes in relative prices and in the relative costs of production of various types of primary products, thanks to events on international markets and to changes in trade and other industry assistance policies at home and abroad. Additional specific policies affecting the minerals sector have been its special taxation arrangements and rules on access to deposits, with the latter becoming more important following the recent passage of Australian native title ('Mabo') legislation.

These types of factors will continue to influence Australia's primary sectors in the foreseeable future, with the added uncertainties during the next decade of how developments in the former communist countries affect international markets for primary (and other) products and how the export outlook will be affected by the interaction of trade and environmental policies. The increasing scarcity of environmental resources in the rest of the world suggests that there will be a further shift in comparative advantage towards the Australian primary products sector, but there are some worrying uncertainties associated with the links emerging between trade and environmental policies.

Regardless of developments abroad, Australia could do much to enhance its economic growth — and particularly its primary sectors' contribution to that growth — simply by continuing to remove its own remaining distortionary policies, including not only the long-standing manufacturing protection policies (especially for cars and textiles — see the next chapter) but also the assistance still provided to some of our least-competitive rural industries. But perhaps even more importantly, other areas of microeconomic reform, including transport, communications, ports and especially labour markets and government utilities and corporations, require attention if production costs for exporters are to be reduced to enable them to fully exploit their potential for international competitiveness.

FURTHER READING

See Corden (1984) for an extended presentation of the model used here to analyse the origins of structural change. Sieper (1982) and Smith (1983) review agricultural and mining sector policy formation. Updates on developments are contained in the annual reports of the Industry Commission.

4

POLICIES AFFECTING MANUFACTURING

Bijit Bora and

Richard Pomfret

The development of Australia's manufacturing sector has depended heavily on protection for its growth. Indeed, a primary aim of Australia's trade policies after Federation (and of some colonies' policies in the nineteenth century) was to diversify the economy by reducing the dominance of trade in primary products (chapter 2). This development strategy has been dramatically reversed by unilateral trade liberalisation, which began in 1973. As the previous chapter showed, the trade policy change coincided with a resource boom, which would have put pressure on the manufacturing sector even in the absence of any policy actions.

This chapter analyses the experience of the manufacturing sector since the trade liberalisation process began. One striking feature of the liberalisation program has been that, while average rates of protection have tumbled, two branches of manufacturing enjoyed increased rates of assistance during the 1970s and continue to be heavily protected (table 4.1). These branches, passenger motor vehicles (PMV) and textiles, clothing and footwear (TCF), are dealt with in greater detail in the third and fourth sections below. During the 1990s there have also been some suggestions that certain activities deserve support if Australia is to establish new export industries. The final section analyses the arguments for assisting sunrise industries.

Table 4.1 Average effective rates of assistance to Australia's manufacturing subsectors, 1977–2000

Industry	*1977–78*	*1982–83*	*1987–88*	*1990–91*	*2000*
Food, beverages & tobacco	10.0	7.0	6.7	3.1	2.0
Textiles	47.0	68.0	65.0	68.0	22.0
Clothing & footwear	141.0	192.0	174.0	176.0	50.0
Wood, wood products & furniture	18.0	14.0	18.0	13.0	4.0
Paper, paper products, printing & publishing	24.0	16.0	15.0	7.0	2.0
Chemical, petroleum & coal products	19.0	12.0	12.0	10.0	4.0
Non-metallic mineral products	5.0	4.0	3.7	3.0	1.0
Basic metal products	10.0	9.0	7.8	7.9	4.0
Basic steel products	NA	14.0	13.0	11.0	6.0
Fabricated metal products	30.0	30.0	22.0	17.0	5.0
Transport equipment	48.0	61.0	46.0	33.0	12.0
Motor vehicles & parts	NA	123.0	88.0	60.0	25.0
Other machinery & equipment	20.0	21.0	22.0	15.0	4.0
Misc. manufacturing	30.0	25.0	26.0	20.0	6.0
TOTAL MANUF.	**23.0**	**21.0**	**19.0**	**15.0**	**5.0**

Source: Industries Assistance Commission, *Annual Report* (various issues); Industry Commission, *Annual Report 1990–91* and *The 1991 Industry Statement*.

Overview of the Australian manufacturing sector

Australia's industry structure has been changing rapidly since the late 1970s. One significant change has been the role played by the manufacturing sector (see chapter 1). Its share of real gross domestic product (GDP) has fallen from twenty per cent in 1976 to less than fifteen per cent in 1992. Between 1982 and 1990 the manufacturing sector reduced its workforce by 73,000 people; output per employee grew at an average rate of 2.77 per cent per year.[1] Figure 4.1 shows quarterly figures of seasonally adjusted manufacturing GDP calculated at constant prices.

Figure 4.1 Manufacturing GDP by subsector, Australia, 1977–92

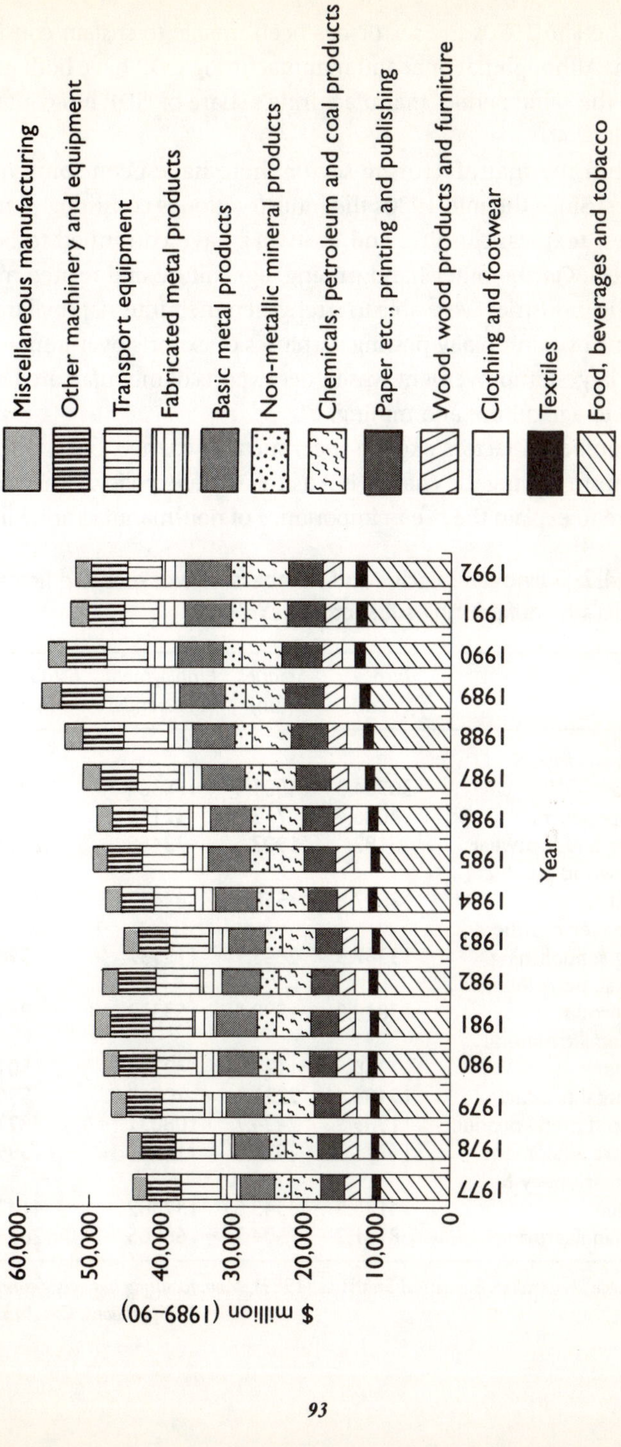

Source: Australian Bureau of Statistics, *Manufacturing Industry: Summary of Operations*, Cat. No. 8202.

Since the late 1970s the sector has been unable to sustain consistent growth. Although national and manufacturing GDP have both grown during the same period, manufacturing's share of GDP has continued to decline.

Within the manufacturing sector there have been some minor changes. Since the mid 1980s the industry groups consisting of motor vehicles, textiles, clothing and footwear have continued to be net importers. On the other hand mining, agriculture and related manufacturing industries were able to strengthen their international trading positions by perennially posting surpluses of exports over imports. By far the largest improvement was in net exports of manufactured goods related to agriculture and mining.

A number of factors explain the structural change, but a principal mechanism driving the change is Australia's trade policy. Other factors do more to explain the rise in importance of non-manufacturing indus-

Table 4.2 Turnover, salaries and numbers of workers and firms in Australia's manufacturing subsectors, 1989

Industry	*Turnover ($m)*	*Salaries ($m)*	*Employment (thousands)*	*Establishments (number)*
Food, beverages & tobacco	30704.3	4001.3	176364	3735
Textile products	4086.4	713.8	32135	717
Clothing and footwear	5182	1303	73488	2311
Wood, wood products & furniture	7897.8	1658.1	84437	4849
Paper, paper products, printing & publishing	13367.3	2863.7	112137	3398
Chemical, petroleum & coal products	13848.8	1590.5	53373	949
Non-metallic mineral products	7268.5	1136.1	42332	1501
Basic metal products	19408.4	2221.1	70745	595
Fabricated metal products	11989	2434.7	108032	4827
Transport equipment	14771.2	2828.8	116285	1599
Other machinery & equipment	14135.5	3343.8	137382	4142
Misc. manufacturing	8141.2	1504.3	66065	2626

Source: Australian Bureau of Statistics (1989), *Manufacturing Industry: Summary of Operations*, Cat. No. 8202.

tries such as financial services and mining than the decline of manufacturing. In the former case Australia during the 1980s had deregulated its financial sector, which accounted for unprecedented growth. In mining many new discoveries and technological advances have led to an increase in investment from foreign and domestic investors in turn increasing production and exports (see chapter 3). Nevertheless, the bold policy moves to reform the manufacturing sector during the 1980s might be having a much larger than anticipated impact.

Table 4.2 provides a snapshot of the Australian manufacturing industry in 1989. Data on turnover, salaries, employment and number of establishments are provided. Australia's comparative advantage in manufacturing based on primary products is clear. Downstream manufacturing of agriculture and mineral products accounts for more than seventy per cent of the manufacturing industry's turnover, employment and establishments. The importance of the transport industry is also highlighted in the table by its contribution to each of the indicators. Textile, clothing and footwear, on the other hand, make a small contribution to manufacturing. This point is interesting to note since in the next section we emphasise the fact that this is the only sector to have received a substantial increase in protection since the 1970s.

MEASURING THE RATE OF ASSISTANCE

There are a number of trade and industry instruments available to a government to support its domestic producers such as tariffs, subsidies and quotas. One feature of Australian industrial policy has been the proliferation of a number of instruments in specific sectors, which creates a problem when it comes to measuring the degree of distortion. Australia currently makes use of tariffs, quotas, subsidies and excise taxes. If only one particular instrument, say a tariff or a subsidy, was used, measuring the impact of the policy in terms of welfare effects and changes in the pattern and volume of trade would be straightforward (see the introduction). However, if within a broad category some policies are not transparent, an accurate estimation of total costs to society will not be possible. For example, an export subsidy would be transparent and its effect on international trade and the level of assistance given to a domestic producer could be easily calculated. On the other hand, a research and development subsidy would be difficult to measure since the outcome of the R&D process is not known. Hence, the value of the subsidy to the producer and the net cost to society cannot be measured. In order to circumvent problems such as these a comprehensive measure is required.

The measures used by the Industry Commission are the nominal and effective rate of assistance.[2] They are based on the concept that assistance distorts market prices. The measures include assistance provided by a number of policies, such as tariffs, quotas, local content schemes and various pricing arrangements, and are based on an estimate of the price of a good in the absence of any distortions. Despite such an extensive list the measures are not comprehensive because of the transparency issue we highlighted above. Nevertheless, the measures are an excellent indicator of the incentives given to domestic producers in protected sectors to alter their production plans.

Table 4.1 shows the average effective rates of assistance in manufacturing industries at the two-digit level of industrial classification. There are two important points to note from this table: first, the decline in assistance given to manufacturing from 1977 to 1992, which is projected to continue to the year 2000; secondly, the high rates of assistance given to textiles, clothing and footwear, and to motor vehicles and parts.[3]

An important theorem in economics is that the effects of a tariff can be replicated by using a producer subsidy and a consumer tax.[4] This theorem can be used to provide a simple measure of the distributional impact of protection on consumers, known as the consumer tax equivalent (CTE). The CTE is calculated by using the level of assistance plus any revenue accruing directly to the government as a result of the assistance measure. The subsidy equivalent (SE) is the difference in the value of output when measured using domestic producers' price and output measured at world price.[5] When this is added to tariff revenue the sum is the total additional cost paid by consumers of this product to support their domestic producers, i.e., the consumer tax equivalent.[6] These concepts can also be illustrated by referring to figure 0.5 when the only instrument of protection is a tariff; the SE is given by the area a + b and the CTE is the area a + b + c. As noted above, the CTE measure used by the Industry Commission would be broader since it incorporates all measures of assistance provided to producers.

It is important to note that production typically requires imported as well as domestic inputs. If a final goods sector is protected and that particular sector uses imported intermediate products intensively then the foreign producers of these inputs would benefit from protection. In order to obtain a more accurate measure of the transfer an additional measure known as the net subsidy equivalent (NSE) is used. The distinction between gross and net subsidy is similar to the distinction between nominal and effective assistance.

Table 4.3 provides estimates for the measures we have just discussed. There are three important points to note in this table in addition to the obvious point that the most costly sectors to Australian consumers are clothing and footwear, food, beverages and tobacco as well as transport equipment: first, the large difference in the GSE and NSE values for food, beverages and tobacco; secondly, the small difference between these values for clothing and footwear; and thirdly, the high NSE values for transport equipment and clothing and footwear.

Table 4.3 Gross and net subsidy equivalents and consumer tax equivalents, manufacturing subsectors, 1990–91 ($ million)

Industry	*Gross subsidy equivalent*	*Tax on materials*	*Net subsidy equivalent*	*Consumer tax equivalent*
Food, beverages and tobacco	1319	1054	264	1143
Textiles	708	141	567	287
Clothing and footwear	1599	324	1275	2063
Wood, wood products and furniture	525	185	340	253
Paper, paper products, printing and publishing	464	185	279	54
Chemical, petroleum and coal products	688	221	467	186
Non-metallic mineral products	128	47	81	27
Basic metal products	773	369	404	
Fabricated metal products	957	389	567	220
Transport equipment	2029	932	1097	1583
Other machinery and equipment	1176	467	709	946
Misc. manufacturing	737	267	470	238
Total manufacturing	**11103**	**4581**	**6522**	**7086**

Source: Industry Commission (1992).

PASSENGER MOTOR VEHICLES

An important element to understanding government policy towards this sector is that it consists of many subsectors.[7] These can be aggregated into two general sectors: assembly and components manufacturing. A completely built-up unit (CBU) consists of manufactured components shipped to a plant where the automobile is assembled. The Australian trade policy objective in this sector had traditionally been to have the highest possible level of Australian content in a CBU; automobiles should be as fully manufactured and assembled in Australia as technically possible. Such a policy was not without its costs, and only recently has the Commonwealth Government backed away from such a strong position and liberalised the automotive industry.

Initial support for the industry was provided before World War I in the form of a tariff on car bodies, which led to an increase in the number of locally manufactured bodies. The combination of high transport costs, imperial preferences and this tariff also established the production of chassis in Australia by North American manufacturers. By the end of World War II Ford of Canada and General Motors of the United States had firmly established themselves in Australia.

The immediate post-World War II era also marked the beginning of an important shift in the industry since it was at this time the Commonwealth Government revealed ambitious plans to build a car in Australia. The prime motivation for this policy was not so much the level of local content in the assembled automobiles but to capitalise on national pride from the end of the war and publicise the fact that Australia possessed the capital and technology to produce a vehicle. Vignettes of the first Holden rolling off the line in 1948 still evoke considerable national pride and sentiment. Apart from this emotional reaction, the growing maturity of the industry in Australia had firmly entrenched the attitude that the industry should continue to be protected.

The principal sources of protection during the 1950s were direct import controls and indirect control of components through foreign exchange restrictions. These were abandoned by 1960, not with an end result of a more efficient and globally competitive industry, but rather a consolidated protectionist alliance, which included Australian component manufacturers and vehicle producers. Foreign companies responded to the abolition of import controls by planning on increasing the range of automobiles sold in the Australian market. They were intending to produce these cars (competing with the standard cars produced by Holden, Ford and Chrysler) by taking advantage of their freedom to import components. This threat resulted in a concerted effort to raise protection and the complex 1964 plan to establish a minimum local content level of 95 per cent.

The goal of maintaining production of finished PMVs as well as supporting domestic component manufacturers was doomed to fail for two reasons: first, the restricted time period within which firms were required to achieve the 95 per cent level, and secondly, the emerging importance of product differentiation, in particular the growth in popularity of small Japanese vehicles. These two factors led to the departure of Volkswagen from Australia, talk of Japanese companies establishing Australian subsidiaries and confusion in the industry.

It is also important to note that the failure of the plan had more to do with its administration than with the policy goal. Maximising local content was the government's objective, mandated by the political process. What is important from an economic point of view is the additional cost borne by society to meet this objective. A number of policies could have achieved a similar level of local content. In particular a tariff on imported CBUs and imported components in conjunction with a subsidy to domestic producers could have achieved the same result.

As noted above, a key feature of Australia's trade policy in the PMV industry has been content protection. The Commonwealth Government specifies a minimum level of local content for a particular automobile or a specific portion of the domestic market available to foreign producers. In this section we shall examine the partial equilibrium effects of this scheme.

The effects of a local content scheme are important from a trade perspective because the objective can also be met by using a tariff and subsidy combination. Figure 4.2, based on Vousden (1987, 1990), compares a local content scheme with a tariff. The price facing producers can be considered as an average between the domestic price of the intermediate good and the CIF price of the imported intermediate good (P_f). Suppose the scheme requires k% of the total value of the production of an automobile to be produced locally. Then the average price (P_a) would be

$$P_a = kP^d + (1-k)\,P_f$$

Figure 4.2 Effects of a local content scheme

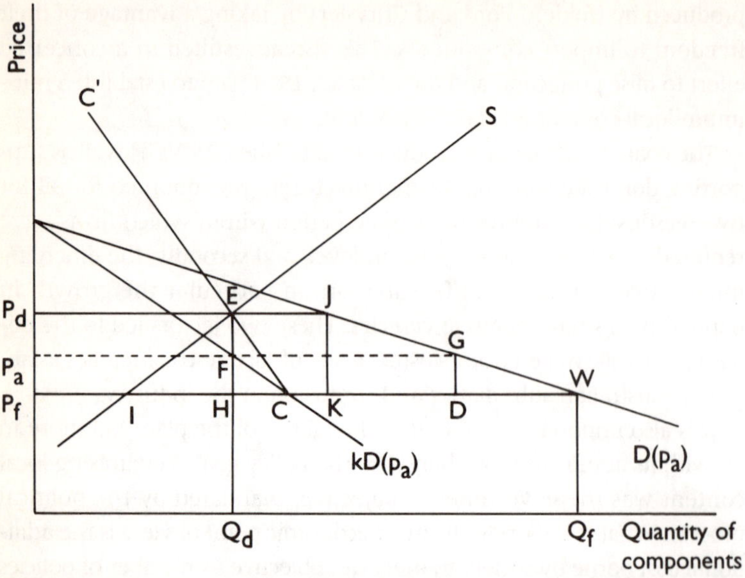

The demand curve with respect to the average price is $D(P_a)$. As with our analysis of market sharing arrangements, the domestic market share is given by $kD(P_a)$. As Vousden (1990) points out, it is important to account for the demand for domestically produced components with respect to domestic price. Consider point C in figure 4.2 where $kD(P_a)$ cuts P_f. At that point $P_d = P_f = P_a$ so it must be one point on the domestic demand curve. Intuitively all other points must be above $kD(P^a)$ resulting in the domestic demand curve (CC'). Domestic supply is given by SS, and its intersection E with the demand curve gives us the domestic price (P_d) and quantity (Q_d). The average price calculated with P_d is P_a, which is read off kD (P_a) at F by drawing a vertical line from E. Total demand for components is given by D(P_a) at point G.

We are now in a position to examine the welfare consequences of a local content scheme (LCS). In the absence of any distortion, Q_f units would be consumed by final good producers at point W. The dead weight loss is, therefore, the area GDW on the consumption side and EIH on the production side. Compared with a tariff of P_f P_d, which would result in the same level of domestic output, the LCS seems to be preferable. As Vousden (1990) points out, the production dead weight loss is the same under both regimes, but the consumption dead weight loss is much higher under the tariff, the difference being JKDG.

In 1973 the Whitlam Government signalled Australia's first shift away from a general protectionist policy to one of liberalisation across all industries. The motor vehicle industry was included in the 25 per cent tariff reduction as well as an increase in the share of the domestic markets awarded to foreign firms from 7.5 per cent to 20 per cent. As documented in Anderson and Garnaut (1987), the Whitlam tariff cut was successful in liberalising all of the manufacturing sectors with the exception of the highly assisted industries. In the motor vehicle industries a continued strong lobby from foreign multinationals and the Australian component manufacturers resulted in the imposition of quotas.

The most important policy change in the industry took place in 1984 when the so-called Button Plan was released by the Minister for Industry, John Button. A key component of the plan was its recognition of the inefficiencies in the industry. Gregory (1987) points out that Australian consumers at the time were willing to pay a 90 per cent premium for the right to purchase an imported vehicle. But if the industry were to be rationalised, a reduction in protection was required, in which case there was every likelihood that the industry might not survive the competition. A mechanism for lowering costs was required to ensure that Australian firms could compete with foreign firms.

The crucial element in the Button cost reduction plan was the role played by increasing returns to scale. Australia is a small country, hence its volume of production was low relative to countries such as the United States. In addition, there was believed to be an over-proliferation of models; in 1984 thirteen different models were produced by six producers. The Button Plan intended to reduce these numbers to six models and three producers by imposing minimum production volumes. Producers were forced to raise production (thereby lowering costs) or to abandon specific models. The policy was a remarkable shift in attitude given the historical treatment of producers. If producers were to remain in Australia, they had to meet these minimum volume restrictions.

Another important element of the Button Plan was the role played by exports. Australian producers had restricted themselves to servicing the Australian market, but if they were to meet the minimum volume restrictions they had to begin exploring overseas markets. As an added incentive the Button Plan allowed firms to accumulate 'export credits', which they could use to offset import duties. In the latest version of the export facilitation (EF) scheme producers are allowed to import up to 15 per cent of their total value of production duty free. If producers were exporting simultaneously they could claim these exports against their 15 per cent.

By allowing exports to count in lieu of local content, the EF scheme is intended to encourage greater rationalisation in component-making and longer production runs for assembled cars. The reaction to the scheme has differed among firms and over time. The initial focus by GM-Holden was on increased specialisation in components. Mitsubishi has also reaped scale economies in component-making by exporting aluminium castings from its Adelaide plant, but Mitsubishi used the export credits to import finished cars and concentrate its Australian assembly operations on the Magna. The most publicised reaction in the late 1980s was Ford's decision to produce the Capri in Australia for export, which led to a rapid growth in car exports (from a low base); in 1992 the Capri accounted for more than three-quarters of Australia's car exports, but the press reports were already switching from the initial praise of a showcase export to concern that Ford might axe the car (which the company decided to do in 1993).

Whether the export credits are used to import components or finished cars is important to the analysis of the costs and benefits of the EF scheme. If a carmaker can import more components than under the basic local content rules, then this must shift down the supply curve and lead to a larger equilibrium output of cars. The re-sourcing of components to lower-cost foreign suppliers is a benefit to the carmakers and to Australia, although the gain is partly offset by loss to domestic component manufacturers. Since the marginal cost of car production in Australia is above the world price, the additional domestic output has a social cost familiar from the analysis of trade barriers in the introduction. There is a second social cost arising from the use of resources to produce exports the marginal costs of which are above the world price; the carmaker will undertake such exports as long as the benefits from access to imports are greater than the loss on exports. In general, the EF scheme must benefit the carmakers or they would not respond to it, but the net social benefit is ambiguous because some of the carmakers' benefits accrue from a redistribution of the rent (and tariff revenue) on imports at concessional duty rates.

The long-term trend appears to be for EF credits to be used increasingly to import finished cars rather than components. In 1989, out of export credits of around $410 million, some $161 million were used to import components and $249 million to import finished cars. The welfare impact depends on whether the imported cars add to or displace domestic output. If domestic output is unchanged, then there is no welfare gain to Australia, but simply a transfer of government tariff revenue to the importer. If, however, the imports substitute for domestic production, then there is a welfare gain from reversing the losses due to protection.

Robert Albon of the Australian National University has made estimates of the benefits and costs of the export facilitation scheme, which give an idea of the orders of magnitude in 1989.[8] The social waste associated with exports is measured by the cost disability, i.e., the percentage by which Australian costs are above the world price (about 10–15 per cent for GM engines and more than 30 per cent for Ford Capris); applying an average of 20 per cent to the $410 million export credits gives a loss of $82 million on producing the exports under the EF scheme. Carmakers must have benefited by more than this amount, and Albon estimates that they saved $64 million on component imports and $106 million on imports of finished cars; the latter is a gain to Australia in proportion to the extent that the imports displaced domestic output, so the $170 million is an upper-bound estimate of social benefit. Against this, specialised component makers lost $14 million, and the EF-induced increase in output involved a social cost of up to $46.5 million. Together with the social cost of $82 million incurred due to increased exports under the scheme, this would yield an upper-bound social cost of about $142 million. The net welfare outcome for society is ambiguous, although Albon believes it is positive because the benefits are sufficiently close to the $170 million upper-bound that they outweigh the $142 million upper-bound estimate of costs.

The welfare calculations compare the situation with and without the EF scheme. As such they suggest that the EF scheme might ameliorate the impact of the rest of the package to protect the PMV industry, but they do not justify the package as a whole or EF in isolation. In particular, the fact that the scheme has promoted Australian car exports is not necessarily a cause for celebration; it might just reinforce the point made in the introduction that Australia could produce almost any manufactured good if it were unconcerned about the opportunity cost. The Ford Capri is a heavily subsidised export because every export dollar earned entitles Ford to save 42.5 per cent duty on a dollar's worth of imported cars. Even with the subsidy, the car flopped in its main overseas market, the USA, never coming close to projected annual sales of 30,000–35,000.[9]

The post-war history of the Australian motor vehicle industry was initiated by a desire for a national car industry on prestige and security grounds, which over time bred a multidimensional policy monster. The 1960s package tried to square the interests of assemblers and component makers by giving the former tariff protection on their output but duty-free import of inputs, while placating the latter with protection through local content rules. By the time of the 1970s trade liberalisation, the PMV industry was large and well organised enough to be able to negotiate exemption. During the 1980s governments tried to resolve the dilemma by incorporating the PMV branch into the tariff reduction program while tailoring programs to prevent the industry's decline.

The cost of these policies has been borne by Australian car-buyers, who are forced to pay well above world prices for new cars and are effectively prohibited from importing used cars (since that would undermine the protective regime). Successive governments have shown little concern for this side of the equation but continue to have faith in their ability to promote rationalisation within the industry. The evidence is, however, that what is needed is a reallocation of resources away from the PMV industry and into activities in which Australia has a comparative advantage.[10]

Textiles, clothing and footwear

The TCF group of industries is the most heavily protected part of the manufacturing sector. In this respect Australia is typical of most high-income countries, and the reasons for the industries' protection are not unique to Australia. Australia does differ from other countries in one key aspect, and that is its role as a net exporter of primary products.

Therefore, Australia's objective is sometimes presented as being to increase the value added derived from downstream processing, although as with minerals (see chapter 3) there is no necessary reason why a country whose comparative advantage lies in producing a raw material will also have a comparative advantage in the related processing activities.

Overall these industries are labour-intensive, and they have faced severe competition from the rapid expansion of low-cost foreign production, especially in East Asia, which has pushed down the world price in real terms. Low-income elasticity of demand and increasing productivity have added to the pressure for the domestic industry to reduce its workforce. The large labour force, geographically concentrated in Victoria and in country towns, has exerted strong political pressure on the government to arrest the industry's decline, and the most obvious approach has been to restrict imports.

The difficulty of competing with low-cost foreign producers has spawned fears that the Australian TCF industries will be eliminated in the absence of trade barriers. This apocalyptic view ignores the heterogeneity of the industries. Although most activities require low skill levels and have high female participation rates, the stereotype does not fit much of the textile branch, which is more capital-intensive than footwear or clothing and employs more men than women. Moreover, niches in all three branches require skilled labour in design or marketing. Thus, although the forces of comparative advantage are working towards a substantial decline in the TCF industry, some activities will survive in Australia.

The TCF industries were the manufacturing activities most affected by the 1973 25 per cent tariff cut because they had the highest existing tariffs and were facing surging imports from the newly industrialising economies of East Asia. Moreover, these industries experienced the largest wage increases (by 91 per cent between 1972 and 1974, when the average increase for manufacturing was 74 per cent), in part because of the Whitlam Government's push towards eliminating gender differentials in wage awards. Under all of these pressures employment in TCF dropped from around 180,000 at the start of the 1970s to less than 120,000 by 1978.

In 1974 a GATT waiver allowed countries to negotiate bilateral voluntary export restraint agreements (VERs), subject to the conditions of the Multi-fibre Arrangement. Following the other high-income countries, Australia initiated negotiations to establish VERs with her major suppliers of textile and clothing imports but, unlike other MFA signatories, Australia soon abandoned the VER approach. The main motive for abandoning the VER framework was that it was insufficiently protective (Cable 1990:120). By including developed as well as developing countries, global quotas prevented undermining of the VER's impact by trade diversion. This hurt Australian consumers and exporters in developed countries but might have favoured suppliers in East Asia, who have typically been most restricted by the MFA. Thus, a secondary motivation behind Australia's withdrawal from the MFA might have been concerned about maintaining cordial trading relations with the Asian exporters, who were also emerging as important markets for Australian primary product exports.[11]

Nevertheless, the government remained sensitive to the TCF industries' demands for protection. The size of the industry was a consideration, but the industry had also developed effective lobbying skills during its long decline. Chance political considerations also played a part. The first by-election contested by Malcolm Fraser as leader of the conservative opposition parties was in 1975 in a Tasmanian electorate centred on Launceston, a city heavily dependent on TCF employment. When Fraser successfully made assistance to the TCF industries a winning component of his election strategy there and again in the 1977 federal election, the industries were assured special treatment from the government.[12]

In 1977 Australia adopted a system of global rather than bilateral quotas, and beyond the quota limit imports were subject to a high tariff rather than prohibition. According to the Industries Assistance Commission about 90 per cent of clothing and footwear and 30 per cent of textile production were then protected by import quotas (Anderson & Garnaut 1987:130–1). Structural adjustment and special assistance funds were also earmarked for the TCF industries. The special measures attracted some investment to the industries but were not enough to arrest its decline. The fundamental problem was that the government was attempting to promote rationalisation *within* the industries to make them internationally competitive, while market forces were pushing overwhelmingly for rationalisation *between* sectors. This was even more difficult in the TCF industries than in the PMV industry, because the former were more heterogeneous with a few integrated multinationals such as Pacific Dunlop alongside thousands of small companies.

The import quotas were based on volume. Their effects can be seen from figure 0.5, which analyses the partial equilibrium consequences of a quota set at KL units. The quota creates excess demand in Australia at the world price P_w, so the domestic price rises until the excess demand is eliminated, at price P_d. The gap between domestic and world price yields a rent on the KL units imported, which is shared between the government (from tariff revenue and any charge for import licences) and import licence holders. Some of this rent, measured by area c in figure 0.5, might be dissipated as people use resources to obtain valuable import licences, but even if that does not happen there will be the usual net welfare costs to Australia, measured by areas b and d in figure 0.5.

The reason why protection remains, despite the net national welfare loss, is that TCF producers benefit from the higher price on the OK units that they produce. Some of this is wasteful because units JK could be obtained at lower resource cost by trading Australian exports for imported TCF items, and some of the quota-induced producer surplus (measured by area a) is dissipated in socially unproductive lobbying to keep the quotas in place. Nevertheless, employees and owners of TCF firms benefit, especially if they possess skills or equipment specific to this industry or if they place a high value on not moving to a job in another industry for personal reasons. The last consideration has generated some public sympathy for the TCF industries because of the large number of married women employed there, but it is important to bear in mind the cost of indiscriminate support for all producers in the industry and the possibility of using more targeted assistance to groups deserving of help.

Moreover, the impact on consumers is regressive, because poor people spend a larger part of their income on these items than rich people. The regressiveness is exacerbated by having volume quotas, which encourage licence holders to import high-value items (e.g., designer shirts instead of cheap no-name shirts), so that it is the price of low-value items that increases the most.

The economic costs are recognised by the government. Under a November 1986 plan, the quotas are being replaced by tariffs, which will be cut to 60 per cent for clothing, 50 per cent for footwear and 45 per cent for textiles by 1996. A generous labour adjustment scheme was designed to soften the blow to TCF workers, and in 1991 an export facilitation scheme (similar to that for PMVs described in the previous section) was introduced to promote rationalisation within the TCF industries. By the early 1990s both major political parties were committed to substantial trade liberalisation in the TCF industries, but the stunning victory in the 1992 Wills (Victoria) by-election of an independent candidate running mainly on the single issue of TCF protection provided a reminder of continuing popular opposition to trade liberalisation in these industries.

SUNRISE INDUSTRIES

Seekers of government support have often pressganged fashionable theories in support of their case. During the nineteenth-century and early twentieth-century debates on protection, proponents directly or indirectly invoked the infant industry argument. During the 1980s one of the hottest topics in international trade theory was the incorporation of imperfect competition into trade models. This resulted in some novel arguments for government intervention, especially in highly concentrated industries where government intervention could influence the strategic interaction among the firms in favour of the domestic firm; the resulting redistribution of monopoly rent could improve national welfare. This argument for intervention became known as strategic trade policy.

Interest in imperfect competition coincided with greater interest in economies of scale. The two often coexist, but can be analysed separately. Because economies of scale internal to the firm are incompatible with perfect competition, the market mechanism might produce an outcome that is not the first-best. Welfare generalisations then become difficult because analysing a policy's impact involves comparing two second-best situations. We would expect scale economies to increase the gains from trade, because they heighten the benefits of specialisation, but cannot rule out the possibility that some interventions might be welfare-increasing when scale economies are present. Protecting the domestic market, so that a firm can cover its fixed costs there and export as long as the world price is above marginal cost, is one example, but here the welfare effects could be positive or negative (and it is important not to confuse increased exports or output of a single product with increased national welfare).

Arguments for pursuing strategic trade policies or intervening in imperfectly competitive industries were heard in Australia in the early 1990s. They appear to have subsided quickly given that the assumed conditions are rare anywhere and especially unlikely in Australia. Few Australian firms are competing in highly concentrated international industries and, where they do, the domestic market is too small for protection to make much difference to the strategic interaction among the firms.

Export subsidies have been advocated for some 'sunrise' industries. The case here appears more based on the perceived success of some East Asian governments in picking winners by judicious intervention and these winners then becoming established firms adding to national welfare. In fact, the evidence on the Japanese and South Korean Governments' success in picking winners is heatedly debated, and Hong Kong, Taiwan and more recently Thailand have experienced more rapid export and GNP growth without significant microeconomic intervention.

Some Australian firms in imperfectly competitive industries have been successful in finding niches and exporting differentiated products. Ken Done and Country Road are examples from an industry in which comparative advantage has moved against Australia; by identifying areas relying on design and other skills, rather than simple production costs, the niche producers have been successful (at least for a while). The Australian film industry and some recording stars are other examples, although whether Mel Gibson or INXS continue living in Australia is crucial to whether their incomes are included in Australian GDP. The point about these examples is that the niches are filled by private initiative, and government officials were not necessary and possibly not competent to sort out potential winners from losers.

In sum, the new trade theories based on imperfect competition and new arguments for intervention developed during the 1980s did enter the Australian policy debates in the early 1990s. Their relevance to Australian conditions is, however, dubious, and they have had no impact on policy.

CONCLUSIONS

Historically the manufacturing sector has been the principal beneficiary of protectionist policies. Since trade liberalisation began in 1973 the across-the-board protection has been greatly reduced. By the mid 1990s the salient characteristic of policies towards the manufacturing sector was their concentration in two branches.

The motor vehicle industry has been governed by a complex range of policies aimed at protecting both component suppliers and car assemblers. The evaluation of the policies can only be understood by recognising the conflicting interests of the two beneficiaries, and successive plans have addressed the weaknesses of the previous policies. The present combination of local content protection and export facilitation provides substantial assistance to both groups, at the expense of Australian car buyers. Moreover, it is far from the least-cost way of supporting component-makers and assemblers, assuming that they deserve to be subsidised.

The textiles, clothing and footwear branch is easier to analyse because Australia's comparative disadvantage in these labour-intensive activities is clear (and increasing) and the policy weapons have been simpler. Protection is costly, and tariffs would be more efficient tools than quotas (and subsidies still better, according to the analysis in the introduction). Reduction in the levels of protection seem inevitable in the coming years, and desirable because the gain to consumers will outweigh the loss to producers. Continuing resistance to the reduction of protection in part reflects public sympathy for the industry's workforce, as well as the regional concentration of the industry, which contrasts with the limited political significance of clothing consumers (everybody buys clothes, but for most voters it is not an item for which they are as aware of paying high prices as, say, cars).

With the exception of these two branches, the political debate on protecting manufacturing industries has almost disappeared. Attempts to marshal theoretical arguments for strategic trade policies in support of intervention in favour of specific activities were unsuccessful and soon abandoned. In the 1993 election both major parties expressed commitment to maintaining trade liberalisation.

FURTHER READING

Anderson and Garnaut (1987) and Anderson (1987a) are the classic works in this area. Boehm (1994) deals with the historical development of Australia's manufacturing sector. Capling and Galligan (1992) is a more recent study on the political economy of industry assistance. The annual reports of the Industry Commission are the best source on current developments.

5

SERVICE SECTOR POLICIES

Christopher Findlay

The service sector accounts for about two-thirds of Australian GDP.[1] It has done so for at least a couple of decades. While it has accounted for a large share of output, many economists used to say that the service sector was not as open as the rest of the economy. They said its component sectors were 'not tradable', that is, they said people did not often trade services across national boundaries.

Contrary to this perception, during the 1980s the service sector became a much more important contributor to Australian exports. Figure 5.1 shows the growth in the absolute value of services exports, which have increased seven-fold since the late 1970s.[2] Their share of national exports (merchandise plus services) has risen to 20 per cent, although this share has changed little since the end of the 1980s (see also table 1.1). The service sector has a smaller export share than it has of output (or GDP). Even so, the service sector accounts for a substantial share of exports. Some categories of services exports, such as tourism or education, are now larger in value than many items of Australia's rural or mineral exports.

This chapter examines two main questions that have been highlighted by the growth in Australian services exports. One question is how services differ from other exports. Services are called 'intangibles'

Figure 5.1 Service exports, Australia, 1978–79 to 1992–93

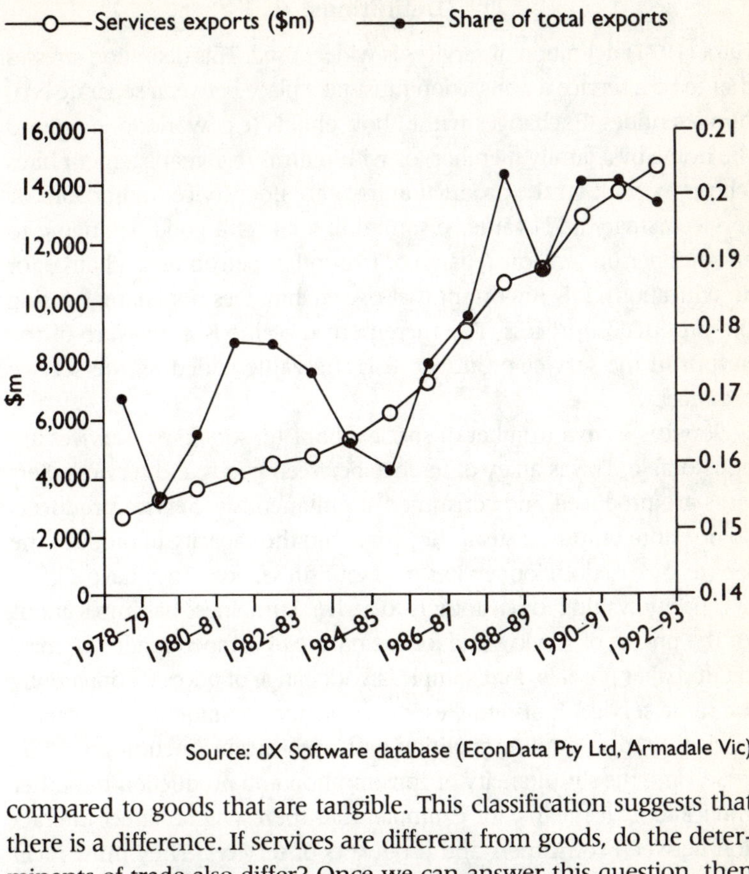

Source: dX Software database (EconData Pty Ltd, Armadale Vic).

compared to goods that are tangible. This classification suggests that there is a difference. If services are different from goods, do the determinants of trade also differ? Once we can answer this question, then we might be able to comment on and anticipate the changing composition of Australian trade and the associated changes in the structure of output of the economy.

The second question concerns government policy on barriers to trade in services. If services are different from goods, do the barriers to trade differ? There might be policies that economists do not usually regard as barriers to trade but that do restrict trade in services. If we can answer this question, then we can suggest ways to increase the gains from trade in services.

THE SERVICE SECTOR

Definitions

Hill's (1977) definition of services is widely used. This definition stresses that to be a service a transaction must take place between separate economic entities. Exchanges within households (e.g. work done around the home by a family member) or within firms (e.g. management hires labour to work on the production line) are not service transactions. A service activity adds value to something, either a good belonging to another person (e.g. car repairs) or to another person (e.g. a haircut or an education). It is important that ownership does not change during the production process. The increment in value is a measure of the output of the service producer. It is the value added by the service producer.

Services have a number of special characteristics. First, services are not storable. This is a key difference between goods and services. Services are produced and consumed simultaneously. Service producers do not store output. Instead they maintain the capacity to produce the service, e.g. of education services or of tourism services. They have a lump of capacity waiting for customers to arrive. Lumpiness has implications for the pricing of services and for the nature of competition in the market. It makes possible, for example, a wide range of prices for apparently the same service. It also makes it possible for incumbent producers of services to meet new competition with vigorous price-cutting.

Second, the simultaneity of consumption and production has other implications. It means, for example, that there will be direct interaction between consumers and producers of services. They must meet in some way in order for the transaction to take place. Direct interaction and the simultaneity of production and consumption also suggests scope for tailoring a service to the demands of a customer, that is, service production is likely to be more highly differentiated than goods production. This feature, combined with the lumpiness in production noted above, suggests that markets for services are more likely to be characterised by behaviour typical of oligopolies or monopolistic competition.

Third, the size of the service sector will depend (as does the size of the manufacturing sector) on choices made about specialisation and trade within an economy. New technologies to which households do not have access, or a rise in the importance of scale economies as a result of the emergence of new technologies, can cause the measured output of services to increase. This happens because new firms appear that specialise in that activity. Changes in the organisation of economic activity can change the relative importance of the service sector as more activities are 'splintered off' and move out-of-house (Bhagwati 1984).

Service sector growth

The Carter diagram (figure 5.2) (see Carter 1987 and Stern and Hoekman 1988) illustrates the evolution of the relative size of the service sector as economic growth occurs. The three sectors of agriculture (A), services (S) and manufacturing (T) are shown at each point of the simplex. A movement away from any point represents a fall in the nominal share of that sector in total output (or GDP). That the service sector becomes important is an interesting phenomenon, and an examination of the reasons provides some useful background information to the discussion of international trade in services.

The observed growth in the size of the service sector is interesting because one analysis of the prospects for the service sector in a growing economy suggested that it might be squeezed. This analysis assumed that the services production process was highly labour-intensive, that labour quality was homogenous and that it involved low levels of productivity growth. Increasing degrees of capital intensity, and hence labour productivity and real wages, in the rest of the economy would raise costs in the service sector. For a given level of demand, and assuming that services are not internationally tradable (see below), this implies rising services prices and falling output (i.e. the service sector supply curve shifts to the left). But, depending on the price and income elasticities of demand for services, the outcome in an economy where incomes were increasing would involve higher services prices but an uncertain effect on volumes traded (the demand curve for services shifts to the right at the same time as the supply curve shifts to the left). In either case, however, we would expect to see a positive correlation between GDP *per capita* and the price of services.

Note that this analysis assumes that there is a given stock of labour in the economy. There is more comment on this aspect of the model later in this chapter.

Figure 5.2 Composition of GDP for selected Pacific economies, 1950–1980

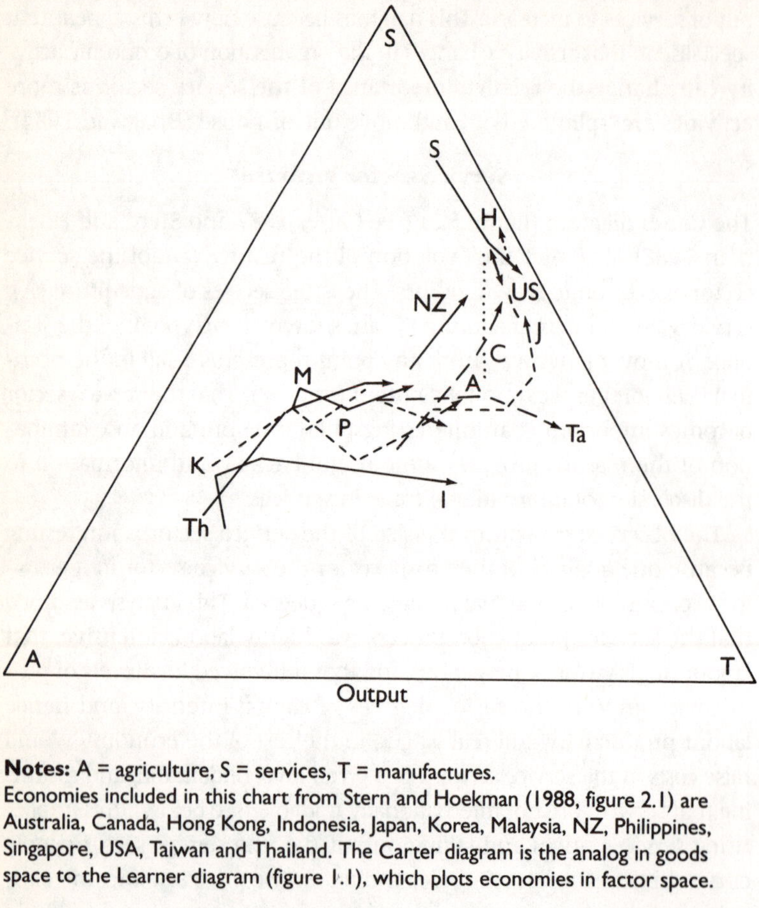

Notes: A = agriculture, S = services, T = manufactures.
Economies included in this chart from Stern and Hoekman (1988, figure 2.1) are Australia, Canada, Hong Kong, Indonesia, Japan, Korea, Malaysia, NZ, Philippines, Singapore, USA, Taiwan and Thailand. The Carter diagram is the analog in goods space to the Learner diagram (figure 1.1), which plots economies in factor space.

Source: Carter (1987); Stern and Hoekman (1988)

The process of adjustment in the service sector might work out in the marketplace, but it might also raise a number of political issues, since many services (e.g. amounting to about 16 per cent of GDP in Australia, or about a quarter of the service sector) are provided by governments. A relatively more rapid rise in their costs will generate a larger government sector and rising tax burden if the volume of services is to be held constant. From this perspective the analysis is pertinent to the concerns about the quality of social infrastructure and the volume of government services provided to the community (e.g. policing, education, health and so on).

Falvey and Gemmel (1989) and Stern and Hoekman (1988) both attempt to isolate the effects of price changes (which they confirm are correlated positively with GDP per head) from volume changes in measuring the size of the service sector. They observe that when the services share is measured in terms of constant prices, that share might not change very much as GDP per head increases. They are measuring the effects of growth and structural change on the volumes of services traded.

These studies refer to the services share of consumption. Measuring services in terms of national output might give a different picture because output measures will include many intermediate services consumed by business. Stern and Hoekman (1988) and Elfring (1989) (the latter only for OECD economies) observe that there is a positive relationship between the service sector share of national output and the level of development (see figure 5.2, which uses output data).

These results are qualified by an important measurement problem, which can lead to an understatement of the size of the service sector. It is difficult to measure value added by the service sector, and instead the contribution of the service sector is measured by the inputs used. For example, the contribution of public administration and other government services not sold in the marketplace is measured by the costs of their production. The value of output could then be severely understated. This method of measurement also makes it very difficult to assess productivity growth in the service sector. If the money value of output is deflated by an index of wages productivity growth is assumed away. This problem might have contributed to pessimism about service sector productivity growth.

There is evidence that as GDP *per capita* rises, the share of the service sector in employment rises. Stern and Hoekman (1988) attribute this both to lower labour productivity growth in the service sector and to changes in economic organisation. However, in a re-evaluation of US experience, Inman (1985) concluded that in the previous two decades the 'splintering off' process has become a much more important determinant of service sector employment growth.

The model in which the service sector is squeezed as an economy grows assumes that the total supply of labour is fixed. In fact, there is a strong correlation between female participation rates and the growth of service sector employment (Stern & Hoekman 1988). It is, however, difficult to identify the direction of causation. The entry of women in to the workforce could also have increased the demand for services that can substitute for work previously done within the family. Falvey and Gemmel (1989) argue that an increase in the supply of labour, either through population growth or higher participation rates, including those of women, can offset the positive correlation between GDP per head and the price of services.

TRADE IN SERVICES

Definitions

There is a large number of typologies of trade in services. The profusion of typologies is related to the simultaneity of consumption and production of services. This requires some sort of contact between buyer and seller, which might take the form of temporary movement of one party or the other or contact by other means (e.g. telecommunications).

Stern and Hoekman (1988) follow Sampson and Snape (1985) and distinguish four types of services trades:

- no movement of providers or recipients
- movement of providers only
- movement of recipients only
- movement of both.

The first type of service transaction has also been called separated services. An example would be the processing offshore of some data set the ownership of which does not change (e.g. credit card companies sometimes have their card slips keyed into a computer in just one location for the whole world; another example would be offshore banking where the account of an Australian resident is managed by a Malaysian bank). Trade in services then is defined as all four types of transaction, as long as residents receive payments from non-residents. The latter point is important. Transactions between residents, even if they take place outside Australia (e.g. a resident buys a ticket on an Australian airline and flies from Singapore to Bangkok), are not instances of international trade. While trade in services can be defined in this fashion, it is actually difficult to collect data on trade in services according to these classifications. Many transactions could be missed as non-residents are not distinguished from local customers, for example.[3]

Other definitions of trade in services are much more narrow. Lee and Naya (1988) include only sales to foreigners from home base. All income from sales made from an offshore base is excluded from 'trade in services'. The rationale for this definition is provided by Arndt (1989).

A country's exports of manufactures consist of that part of the output of its manufacturing firms that is sold to non-residents. By analogy, one would expect a country's exports of services to be defined as that part of the output of its service industry firms that is sold to non-residents.

Arndt stresses that it is important to distinguish between factor and non-factor services. If a firm sets up offshore, even in the service sector, then that is foreign investment and should not be counted as a service transaction (contrary to Sampson and Snape's classification). Arndt would permit the temporary movement of a firm to the site of the service consumer to be counted as a service trade (it takes place from a domestic base), but not if that firm set up offshore and sold services from that base, and not if its employees 'moonlighted' and sold their services independently of their employer from home. Therefore, in Arndt's system, incomes associated with factor flows are not services transactions. This is certainly consistent with the point made about the definition of a service and the exclusion of transactions such as those within firms.

The key difference between these two views is the treatment of the case where the providers move to a permanent offshore base. The argument in favour of making the distinction (the Arndt position) is that the analytical apparatus differs between the explanation of trade and of offshore investment. The determinants of patterns of trade from home base are different from those explaining the direction and form of factor flows. This has important implications for the collection and analysis of the data on trade in services (see below).

The argument in favour of the inclusive definition (the Sampson and Snape position) is that trade and investment flows are substitutes and inevitably are thought about at the same time in the policy debates.

Thus both views have a role to play. In the following sections, this chapter stresses the Arndt position when attempting to interpret the Australian data on trade in services but takes the Sampson and Snape position when discussing issues in services sector trade policy.

Characteristics of Australian trade in services

While the service sector has some distinct differences compared to goods in the transactions, it is still involved in a value-adding process that consumes the factors of production. It can be expected, therefore, that the factor intensity of the service production process, relative to the factor endowment of an economy, will affect the international competitiveness of that economy in services production. A country well endowed with factors used intensively in services production would be expected to have larger net exports of services (from home base) and have a higher share of services in exports than on average.

The share of services in world trade has been remarkably steady at about 20 per cent.[4] It was at this level in 1970 and was still at this level in the early 1990s. Hirsch (1989) interprets this figure, which is lower than the share of services in GDP, as evidence of the constraints on the tradability of services because of the simultaneity factor. On the other hand, it might also have been expected that the shares of services in world trade could have risen over this period because of technological or policy changes that made services more internationally tradable, e.g. jet aircraft and their impact on tourism, changes in regulatory policies and their impact on financial services, and changes in information technology.

The share of services exports in Australian total exports is also about 20 per cent. A country could be said to have a revealed comparative advantage in trade in services if its share in that country's exports exceeded its share in world trade. The data for Australia suggests a ratio of one. According to this ratio, Australia is about average and is neither an especially competitive or uncompetitive services exporter. What are the prospects for services exports from Australia?

There appears to be no stable relationship between the services share of exports and GDP *per capita*. One hypothesis is that, at low levels of income, growth in GDP *per capita* increases the contribution of the service sector to GDP but not to its importance in exports. There is a correlation between GDP *per capita* and net services export for middle-income countries but, once industrialised economy status is attained, the correlation disappears again. Stern and Hoekman (1988) conclude that the industrialised countries are not necessarily the most dynamic service exporters.

This result is consistent with the earlier analysis, which suggests that, at the aggregate level, the domestic costs of services in general tend to rise with income levels. Comparative advantage in the service sector might be lost at higher levels of income (the opportunity cost of services sector output increases). Instead of prices rising, an outcome where services are internationally traded might be that the import share of consumption rises, domestic prices rise by less and services consumption grows more rapidly. However, the simultaneity factor inhibits trade from the home base of the provider and creates some natural protection for the local service sector. Foreign entry through setting up a local base (i.e. direct foreign investment) is discussed below.

Some interesting data on the issue of when a country is expected to be a more or less competitive exporter of services are reported by Toh and Low (1989). They suggest that on average in Singapore the service sector had a higher ratio of employment per unit of fixed assets than did manufacturing, but it also had a higher level of value added per worker. This suggests that the service sector is on average relatively skill-intensive. It will tend to be competitive in countries with more high-quality human resources compared with other factors. It might be especially competitive in such economies at early stages of development when real wages are relatively low. The relevance of this stages perspective on competitiveness in the service sector is also stressed by Sapir (1985).

There are two views on the significance of the empirical result of the lack of correlation between stage of development and a services orientation in exports. First, recent Australian services export growth reflects a shift in the competitiveness of the sector, which is evidence of the sorts of structural changes that have occurred in Australia in a relatively short period. Furthermore, it suggests that the service sector is recovering its international competitiveness, which is important for the other business consumers of services. Second, the fact that services trades are not dominated by the industrialised countries is important evidence in the debate about liberalising services trade, which is discussed below. Developing economies also stand to gain from more liberal trading rules for services.

There are also qualifications to the result that the industrialised countries might not be competitive services exporters, such as the possibility of an increase in the labour supply and of productivity growth across the board in the service sector. Also some traded services use factors of production other than labour and capital, and some have more or less scope for productivity gain. An analysis that takes account of the endowment of those specific factors, and of differences between the subsectors of the services group, will be more powerful. For example, Australia is expected to have a comparative advantage in the provision of land or natural-resource-intensive services, like tourism or other forms of personal services where space is important (learning to fly a plane or drive a car!).[5]

The patterns of trade will differ between components of the sector. However, it is difficult to test propositions about the links between trade and factor intensities because of the aggregate nature of the available data.

Figure 5.3 shows the change in the composition of Australian services exports since the late 1970s. The categories used are the most detailed

Figure 5.3 Composition of services credits, Australia, 1978–79 to 1992–93

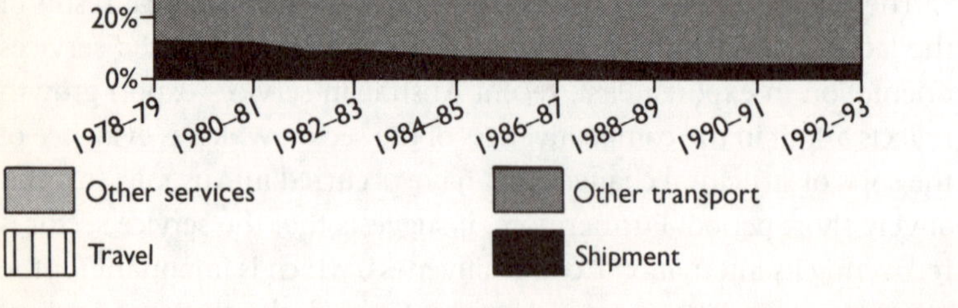

Source: dX Software database (EconData Pty Ltd, Armadale Vic).

available, namely shipping (charges paid to Australian shipping companies), transport (fares paid by travellers to Australian companies), travel (other purchases by visitors) and other services. The 'other services' category includes items such as services to businesses (accountancy, legal services, financial services, advertising and so on). The major change over this period has been the growth in tourism exports. Australia is not a consistent net exporter of any of these categories. For example, combining travel and other transport gives an estimate of the net trade in tourism services. This series is shown in figure 5.4. There has been a shift in the tourism category from persistent net imports towards net exports. Its share in total exports has also been rising. The tourism sector is Australia's largest services export income earner.

Other services have been a consistent net import item in Australia. That is also the case for shipping (not shown). Australia is a net importer of services in total (see figure 5.5). However, since the mid 1980s, the net export ratio (net exports relative to the total value of exports and

Figure 5.4 Net tourism and other services exports, Australia, 1978–79 to 1992–3

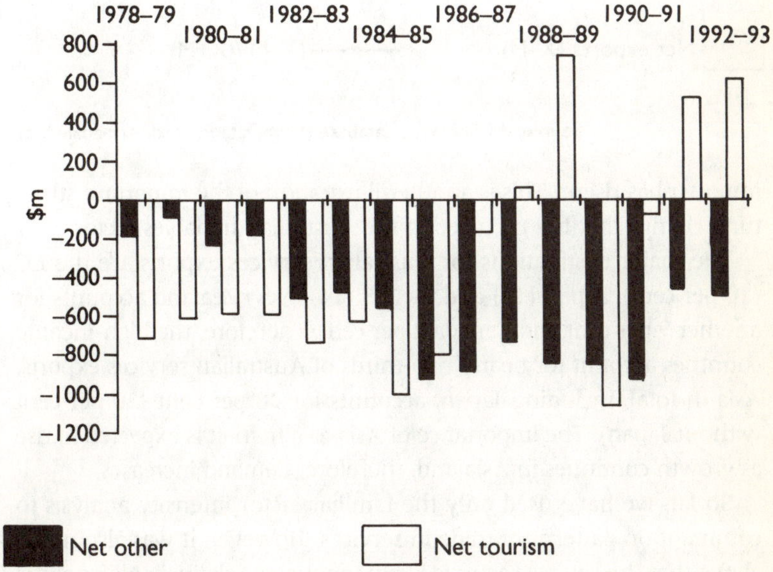

Source: dX Software database (EconData Pty Ltd, Armadale Vic).

Figure 5.5 Net services exports in $m and as a share of total services trade, 1978–79 to 1992–93

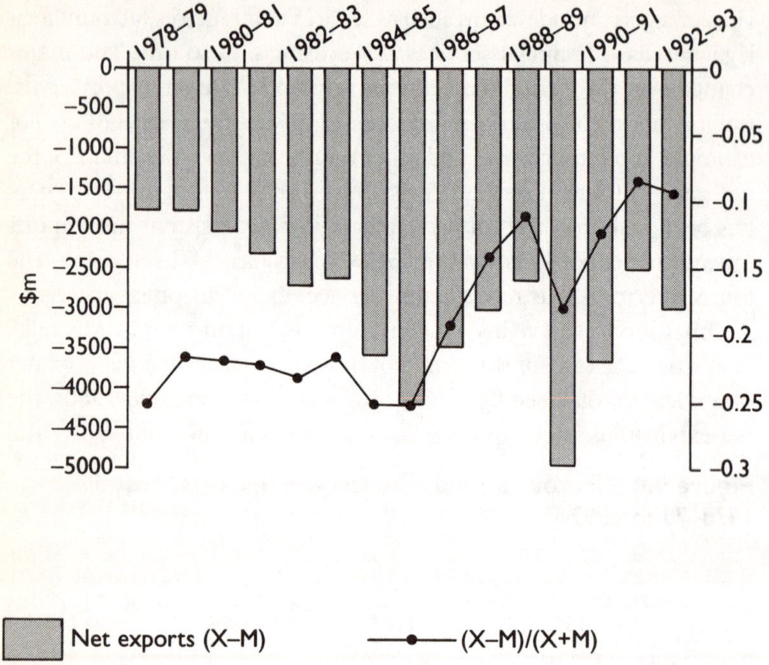

Source: dX Software database (EconData Pty Ltd, Armadale Vic).

imports) has risen.[6] This is another illustration of the important structural change that has occurred in the Australian tradables sector.

The major destinations for Australian services exports are the EC (20 per cent), Japan (18) and the US (14). New Zealand accounts for another 9 per cent and Canada 2 per cent. Therefore, the high-income countries account for nearly two-thirds of Australian services exports. Asia in total, including Japan, accounts for 38 per cent (20 per cent without Japan). The importance of Asia as a market is expected to rise as growth continues in Asia and, therefore, demand increases.

So far, we have used only the familiar factor intensity analysis to comment on patterns of trade in services. However, it was also noted above that the service sector is likely to involve relatively oligopolistic markets. This suggests that more recent theories of trade patterns that stress imperfect competition might be highly relevant to explaining patterns of trade in services. One reason there has been little development of this idea is the difficulty of obtaining relevant and relatively disaggregated data on that trade.

Foreign investment in services

While the service sector accounts for about 20 per cent of world trade, it has been estimated to account for about 50 per cent of direct foreign investment flows. Contrary to its constant share of world trade, the service sector share of the total world direct foreign investment (DFI) flow is much higher than that of the 1970s. There has been a relatively more rapid growth of DFI in the services sector than in manufacturing. Thus, DFI is relatively more important than trade from a home base for the service sector than it is for manufacturing.

Much of the growth in services DFI has taken place between developed economies. It tends to be dominated by financial sector investments and trade-related services. In some economies it is also substantial in other sectors; e.g. tourism and construction as in Australia, and transport. There are some highly visible multinational corporations in the service sector, such as fast food (McDonalds), accounting (KPMG Peat Marwick) and advertising (Young & Rubicam) firms, but these are generally not the most important activities in terms of the total DFI flows.

Figure 5.6 shows the composition of DFI flows into Australia. Consistent with world data, the service sector accounts for about half the total. Much of the service sector inflow, however, is in finance and property. In this respect, Australia might be an outlier.

Overview

Australian services exports grew rapidly during the 1980s, so much so that the services share of total exports rose to 20 per cent. A high-income country like Australia is not necessarily expected to have a comparative advantage in service activities. Indeed, the recent growth in services exports only restores their share of Australian exports to the average in world trade. Overall Australia is a net importer of services traded from the home base of the provider, but net imports as a proportion of total services trade have been declining.

There are important differences between factor intensities in services production at less aggregated levels. The growth in services exports is

Figure 5.6 Composition of foreign direct investment, Australia, 1985–90

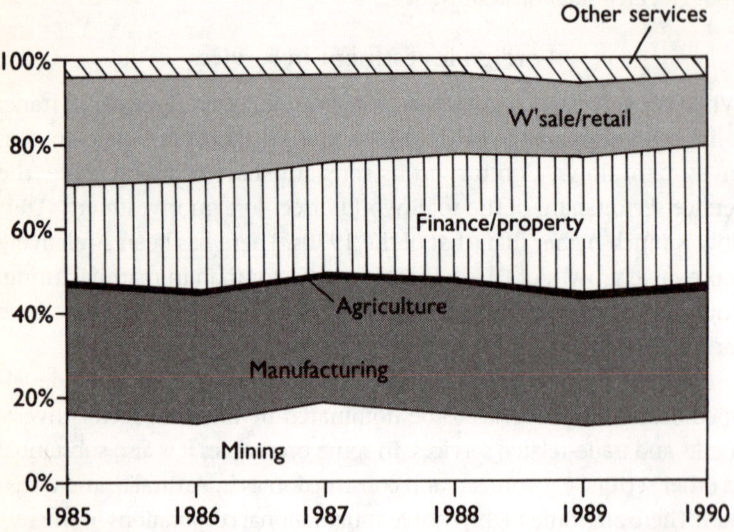

Source: Bora (1992)

associated with a rise in tourism exports. Australia could be expected to have a comparative advantage in a relatively space-intensive activity like tourism.

The importance of the services sector in DFI flows is large relative to its share of trade. Australia might be an outlier in the high concentration of that investment in property and financial services.

POLICY ISSUES

There is a variety of barriers to trade in services. They include subsidies, both direct and implicit, in the operation of government-owned firms, rationing of foreign exchange funds, local content rules, barriers to entry into the market or to local infrastructure required to service the market and rules on the establishment of local operations.

It is important to note that the list does not include tariffs. It would be hard to impose a tariff on a services import. The impost would have to apply to the resident, or the good belonging to a resident, which moves across the national border and which consumes the service of a foreign provider. If the foreign producer moved here to supply a resident, then the tax would have to apply to the capacity of that producer to supply services. The design and implementation of a tariff system for services is not as easy as it would be for goods. Not surprisingly, then, barriers to services trades tend to be various types of quantitative restrictions such as those listed above.

The inclusive definition of services trades (i.e. that of Sampson and Snape) included categories when demanders and providers move. The barriers to trade where demanders move are likely to be less significant than those where the providers move. Important barriers in the former case will be restrictions on travel (visa restrictions) or restrictions on the use of foreign exchange imposed by the home country. A services exporter might also tax its services export sector and reduce the volume of exports, e.g., by asking visitors to make transactions at a relatively high value of the local currency.

In summary, not only does it appear that sales from foreign base is much more important than sales from home base for service providers (see previous section), but also the barriers to sales from a foreign base are also more significant and demand more attention and policy debate. It has been the relatively more rapid growth of DFI in services that has promoted the discussion of policy issues in the service sector.

Barriers to services trade are usually embodied in domestic regulation. That regulation might directly exclude foreign suppliers from a market or might do so indirectly through its other requirements (e.g. meeting particular standards that serve to discriminate against foreigners). Australian regulation of the financial sector, air transport, telecommunications and shipping all have the effect of either excluding foreign suppliers outright or issuing licences for foreign entry and in many cases imposing quotas on the capacity they can offer.

An extra complication in the Australian case is that those domestic regulations are often in the hands of state governments. Reform of rules of market access and on the performance of other suppliers in the market, such as government business enterprises, are also important and in many markets are in the control of state governments.[7]

It is difficult to estimate the distortion imposed on consumer prices as a result of services sector regulation. There have been some significant examples of unilateral liberalisation in the service sector in Australia, e.g. the entry of foreign banks, the integration of the New Zealand and Australian air transport markets, and the entry of a competitor into the telecommunications market. However, it is possible that service sector distortions are still far more significant than those remaining in the manufacturing sector. There have also been examples of regulatory reform in domestic services markets that have fallen short of some degree of liberalisation. There is an important distinction between deregulation and liberalisation. The latter involves foreign entry whereas the former does not: it might, however, involve a greater degree of competition between domestic suppliers. Deregulation is usually the first step, but the same forces that encourage deregulation might ultimately lead to liberalisation.

The origins of the pressures for deregulation and liberalisation in the services sector in Australia are interesting to consider. As documented in other chapters in this volume, the levels of assistance for other tradable sectors has been falling. The decline in protection and a commitment to not reversing that decision redirected the lobbying effort in the tradables sector towards achieving lower costs of inputs. That meant attacking regulation and trade barriers that affect the costs of procuring business services. Business users tend to be more effective in the political process than consumer groups and are more likely to affect the regulatory outcome. The result has been pressure for regulatory reform in the service sector, which has also had implications for Australian trade in services.

The GATT and regional approaches to liberalisation

The previous section noted the efforts to liberalise trade in services in Australia on a unilateral basis. Australia has also been involved in multilateral attempts to liberalise the trade and is likely to become involved in regional attempts to do so as well.

There has been substantial effort during the Uruguay Round to apply the GATT principles to services trade. The rapid growth of world trade in services has highlighted the issue. Some of the important GATT principles are transparency, national treatment and most favoured nation treatment (see chapter 9).

Transparency is difficult in the services trades because of the nature of the international transaction. The usual method of raising the level of transparency is to convert other trade barriers to tariffs. As noted already, it is difficult to see how this could apply in many services trades. Negotiations on market access are concerned with the reduction of these barriers. National treatment means that all trade barriers should apply at the border and there should not be another source of discrimination between local supplies and imports. The most favoured nation (MFN) principle means that there should be no discrimination between foreign suppliers.

These principles will be applied in a General Agreement on Trade in Services (GATS). This comprises a 'framework agreement' on how the principles apply to services. Reflecting the interest in DFI in services, the GATS includes in the definition of trade in services the presence of foreign service providers. In principle, therefore, the GATT rules also apply to all modes of delivery, including MFN and transparency. The GATS includes a number of annexes. One is a list of exemptions that countries claim from the MFN principle. There are annexes on three service sectors — air transport, financial services and telecommunications — that provide a modified application of GATT rules to those services. The GATS will cement the status quo, and there will be further rounds of negotiations to reduce trade barriers.

The GATT applies globally to all members. Smaller groups of countries could agree on trading rules for their region. The service sector is included in the NAFTA agreement between the US, Canada and Mexico, for example. Also, the APEC process[8] is considering options for liberalising trade in the Asia–Pacific region, and some services trades could be included.[9] Regional negotiations, involving with a smaller group of economies, could be effective on service sector trades that have not been liberalised in the GATT process. For example, a regional group might agree on a rules for foreign investment in services if the GATT has to sidestep that issue.

Given the intensity of trade between the negotiating countries and, therefore, the concentration of the benefits of liberalisation among them, the negotiators in a regional grouping might be prepared to carry the free-riding behaviour of other economies not involved in the regional agreement. This presumes that the outcome is not discriminatory. The presence of the GATT framework agreement will act as an important constraint and reference point for those regional talks.

CASE STUDY: TRADE IN EDUCATION SERVICES

The education sector has become one of the largest service sector export industries as well as one of the largest of all Australian export earners. Gross revenue from education exports in 1993 was estimated to be $1.6 billion, which was larger than the value of exports of (for example) iron and steel or aluminium.

There are some interactions between the domestic and export markets for education services. The student population in Australia includes Australian undergraduates who pay the Higher Education Contribution Scheme (HECS) fee, Australian postgraduates, some of whom have fee scholarships and some of whom pay full fees, and foreign students who pay full fees. In 1992 the total number of places was about 559,000, an increase of 42 per cent over 1987. The development of the market for overseas students has been reviewed by Harris and Jarrett (1990).

The case study illustrates the interaction between regulation of the domestic market and export volumes. It also provides an example of incentives for offshore relocation of service sector providers.

These two markets are illustrated in figure 5.7. In that figure, D_A is the Australian demand for tertiary education. For the sake of this modelling exercise all places in tertiary institutions are assumed to be identical. These units (or places) are shown on the horizontal axis. S is the supply curve for tertiary places in Australia. The domestic market will be examined first, then the export market will be introduced into the model. Australian students also travel overseas to purchase education services (that is, Australia is also an importer of education services) but for the sake of simplicity that form of the trade is ignored here.

The domestic market for tertiary places is highly regulated. The Commonwealth has established a 'global stock' of places at each university as well as the expected growth of each university. Of the basis of these assessments, the university is provided with funds to supply places to Australian students. In terms of the static analysis in figure 5.7, this means the government has set the total number of tertiary places to be made available to Australian students. This volume is shown as OQ^* in figure 5.7.

How are these places allocated? If left to the market, the stock of places would be allocated at a price of OG'. All those people located

Figure 5.7 The market for tertiary education services

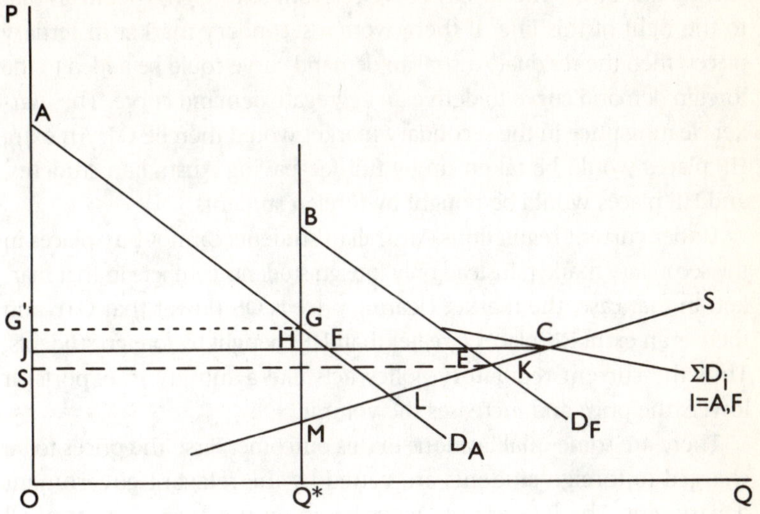

along the demand curve between A and G would buy a place. Places would be allocated to those most willing to pay. There is then a residual demand curve, shown by D_A to the right of point G.

In fact, places in tertiary institutions are not allocated on the basis of willingness to pay. They are allocated on the basis of scores earned at the end of high school. This means that places are allocated to students who do well in those end-of-school exams. The entrants to the tertiary system might therefore not coincide with those who would enter under a willingness-to-pay system. The residual demand curve might therefore lie above that shown by D_A to the right of point G.

Some fees are charged to students who win a place but this fee (HECS) is less than 40 per cent of estimated full costs. This fee is less than the market clearing price OG', and at that HECS fee there is excess demand for places by Australian students. The excess demand has been estimated to be in a wide range from 30,000 to 80,000 students, that is, from 22 per cent to 37 per cent of the available number of places in 1992 (but lower in 1993).

The export market, on the other hand, is not so highly regulated. In figure 5.7 the demand for places in Australian institutions by foreign students is shown as D_F. This curve is drawn with respect to the vertical line at Q^*. The supply curve relevant to foreign students is that to the right of this line. If there were a secondary market in tertiary places, then the residual Australian demand curve could be added to the foreign demand curve to derive an aggregate demand curve. The market clearing price in the secondary market would then be OJ. An extra HF places would be taken up by full-fee-paying Australian students, and HE places would be bought by foreign students.

Under current regulations Australian students cannot buy places in the secondary market. Instead only foreign students compete in that market. In that case, the market clearing price is OS (lower that OJ), and there is an extra RK places (greater than HE) bought by foreign students. Thus the current regulatory policy acts like a subsidy to exports. It lowers the price and increases the volume.

There are some qualifications to this outcome. First, the prices to be charged to foreign students are vetted by the relevant government department. The fees are estimates based on the average costs of all Australian institutions, and they are set as floor prices. If prices based on these averages are set above the market clearing price, the demand by foreign students will fall, and there will be excess supply for foreign students.

Second, the supply curve might have a flatter shape than shown in figure 5.7. If, for example, the supply curve over the relevant range were horizontal, then the export price would be independent of conditions in the market for domestic students, and the degree of interaction between the two markets is eliminated. The shape of the supply curve is therefore a key empirical issue in evaluating policy options in this sector.

Third, the position of the demand curve for foreign students might not be independent of the number of places available for domestic students. There is another regulation that the same cut-off scores be applied to foreign as to domestic students. As the level of places available to domestic students increases (Q^* shifts to the right) then the cut-off scores fall. This will make more foreign students eligible for entry so that those who might bid for a place will increase and D_F shifts to the right.

There remains a demand for Australian tertiary places even when the domestic market clears. This is the demand curve D_F to the right of point K. The extent of this residual demand will be even greater if there is binding price control in the export market. Some Australian universities have sought to supply this demand by offering substitute products from offshore locations; that is, and as is possible in the services trades, the suppliers have moved to the demanders in addition to the opposite movement.

Other costs incurred by consumers of education services include those of travel to and accommodation in Australia. These items might sum to an amount even greater than the fees paid. The provision of education services from an offshore base will also be an attractive commercial strategy from the point of view of Australian institutions for this reason. As offshore campuses are established, the demand for services from local institutions (D_F) will shift to the left. Emerging technology relevant to distance education could also reduce the demand by foreign students for courses supplied from the provider's home base.

CONCLUSIONS

This chapter began with two questions. One question was how services differ from goods. The key difference is the simultaneity of consumption and production of services. That means that international exchange in services might require the movement of a service provider or consumer. Direct foreign investment is more important compared to trade from the home base of the provider for services. Services DFI has grown more rapidly relative to that in manufacturing, and this growth in interest in setting up business offshore has prompted the international policy debate in the services sector.

While services differ from goods, the principles of comparative advantage still apply to services traded from a home base. The composition of services transactions will vary according to the factor intensity of the production process and in conjunction with national endowments of factors of production. While the size of the service sector tends to rise with income per head, on average the relative price of services also increases. This feature and the variations within the sector indicate that there is scope for substantial gains from trade in services between economies at different stages of development and with different factor endowments. However, the simultaneity factor provides some natural protection for local service providers. Recent rapid growth in Australian services exports has returned Australia to par in terms of the services share of total exports.

The second question concerned government policy on barriers to trade in services. If services are different from goods, do the barriers to trade differ? There are policies that economists did not usually regard as barriers to trade but do restrict trade in services. Australia has been involved in unilateral and multilateral attempts to reduce barriers to trade in services. There is substantial pressure for policy reform in the service sector. This pressure is related to the relatively more rapid reform in other tradable sectors. In future, Australia is also likely to be involved in regional approaches to liberalisation of trade in services.

FURTHER READING

Industry Commission annual reports provide more detail on barriers to trade in particular service sectors. Carter (1987) provides a longer-run perspective on service sector growth. The literature on the trade is reviewed by Feketekuty (1988) and Findlay (1990).

6

Trade Policies and International Factor Flows

Bijit Bora

The role of foreign capital and labour in the development of the Australian economy has been highlighted by a number of authors.[1] As a small open economy with an abundance of natural resources and a sparse population, foreign capital and labour were required to develop and diversify Australia's industrial base. In fact, as was highlighted in chapter 2, population growth and the need for foreign capital were considered to be synonymous with economic growth. In this chapter we focus on Australia's use of trade policy to host and attract foreign factors of production.

Australia is similar to any other country in the world in that it recognises the contribution of foreign factors but is protective of its national interest. The Foreign Investment Review Board assesses incoming foreign investment, and Australia has always favoured a selective immigration policy. Despite these screening devices Australians still tend to display a certain anxiety to foreign penetration. A recent example was the national controversy that erupted when the American food giant Campbell's attempted to gain a majority share of the Arnott's biscuit company. Arnott's, of course, ranks with the first Holden, meat pies and footy as a truly venerable icon of Australian culture. Protagonists on one side of the debate could not bear the thought of this Australian company under foreign ownership.

Similar arguments seem to surface when the issue of Asian immigration is discussed. Recently concerns have been raised about the desirability of increasing the intake of Asians. This scenario is all too familiar to Australians as they had to answer the question of whether all Caucasians were similar immediately after World War II. At that time Australia's immigration policy did not permit non-whites to emigrate to Australia, but an abundance of displaced persons from Europe were looking for a country to start a new life. Accepting Greeks, Italians and Portuguese immigrants was not an easy task for Australians who had identified themselves closely with and preferred British immigrants.[2]

Issues based on nationalist sentiment are better dealt with elsewhere.[3] Within the pages of an international economics book we shall confine ourselves to the negative and positive contribution of foreign factors of production to the Australian economy. Specifically, we address the issue of the use of trade policy as an instrument of economic development when factors of production are freely mobile across countries.

Our discussions in this chapter will be divided into three sections. First we look at how trade policy affects the flow of foreign factors, and Australia's use of these policies in the past. Chapter 1 contained some discussion of Australia's net international investment position; the second and third sections of the present chapter will disaggregate that analysis and examine one particular component of non-official borrowing, direct foreign investment. In the fourth section we look at the importance of net migration to Australia and its impact on the current account balance. We conclude the chapter with a discussion on the current direction of Australia's foreign investment and immigration policy.

LOCATIONAL DETERMINANTS OF FACTORS OF PRODUCTION

Before we can discuss the effects of trade policy on factor flows it is important to understand the basic locational determinants of factors of production. In the neoclassical framework, factors migrate between countries in an attempt to maximise their earnings. Hence, they flow from regions where their returns are low to regions where they are high. Of course, there are certain types of factor flows whose locational determinants encompass more comprehensive motivational factors. For example, there is a clear distinction between portfolio capital flows and capital invested with the intent to gain control. Another example is migration for family purposes. Australia has, historically, attracted immigrants interested in employment. Today, however, its composition of immigrants has started to change as more and more applicants are being accepted under the family reunion program (Withers 1987; Wooden et al. 1990).[4] By definition immigrants in this category value family links above employment prospects. In both these cases labour and capital flow not in response to pure factor price differentials but in terms of a more composite index of returns.

Despite these complexities economists have derived useful insights from models that generate factor flows in response to factor price differentials. An extensive literature has developed on the effects of trade policies on the returns to factors of production. The principal result in this area is the Stolper–Samuelson theorem, which states that factors used intensively in a protected sector will benefit from protection. The higher returns due to protection will encourage greater inflows of that factor.[5]

From an Australian perspective it is interesting to note that there is some debate about whether this result had been foreseen when Australia's post–World War I tariff system was designed. At that time Australian policymakers wished to develop the country's manufacturing base, and the tariff was seen as an appropriate instrument (see chapter 2). This case for protection was contained in the now famous Brigden Report.[6] The report was important not only for its stance on protection but also for its recognition that protection, migration, foreign investment and economic development were all linked. The question was whether an increase in the flow of foreign factors offsets the cost of protection.

Australia's preoccupation during the inter-war period with economic growth and development resulted in the adoption of many of the recommendations contained in the Brigden Report. In chapter 4 of the present book the motor vehicle industry is cited as a case in which foreign capital in the form of foreign direct investment by General Motors and Ford was obtained by a guarantee of protection. As well, many of Australia's immigrants during the period found work in these automotive plants and the labour-intensive textile, clothing and footwear industries. In some sense the role of protection was consistent with the infant industry argument, but it is important to note that Australia had a complementary policy of open borders to foreign capital and labour. It is doubtful whether Australia could have achieved the current level of diversification in its manufacturing sector without these inputs.

In order to illustrate the effects of protection-induced flow of factors consider the two-sector factor endowment model with full employment (described in the introduction). We shall assume technology is such that good X is labour-intensive and good Y is capital-intensive. If a tariff is imposed on good X its domestic price rises, and it becomes more profitable for some firms to relocate from the Y sector to the X sector. However, in doing so the difference in factor intensities means that less labour will be released from the production of good Y than is required for the increase in production of good X. Similarly, more capital would be released from good Y than can be absorbed in good X. Perfectly competitive factor markets would result in the return to labour rising to clear the labour market and the price of capital falling to clear the excess supply of capital. The end result is that the tariff has increased the return to labour, used intensively in the protected sector, while lowering the return to capital. This result is commonly known as the Stolper–Samuelson theorem.

In figure 6.1 we have drawn a production possibilities frontier (AB) for a country that produces two goods with the same intensities as we have described above. The line P_W represents the world price ratio, and P_T is the tariff-distorted domestic price. The tangency of P_T to AB gives domestic production of each good. Starting from a free trade equilibrium at F, suppose a tariff is imposed on labour-intensive imports. In this case, illustrated in figure 6.1, production of the labour-intensive good increases (production moves from F to C) and by the Stolper–Samuelson theorem the wage rate rises. Foreign labour is interested in capitalising on the higher wage rate flows into the economy. If we assume full employment then point A' on the new production frontier where the economy produces only good X must lie to the right of point A. Similarly, B' where only good Y is produced must lie on the vertical axis above B. Since good X is labour-intensive, point A' must be further away from A than B' is from B. Therefore, the post-immigration production possibilities curve is given as A'B'. And domestic production is given by the tangency of the tariff-distorted domestic price line to the new production possibilities curve at point C'.

Figure 6.1 Effect of factor growth on production possibilities

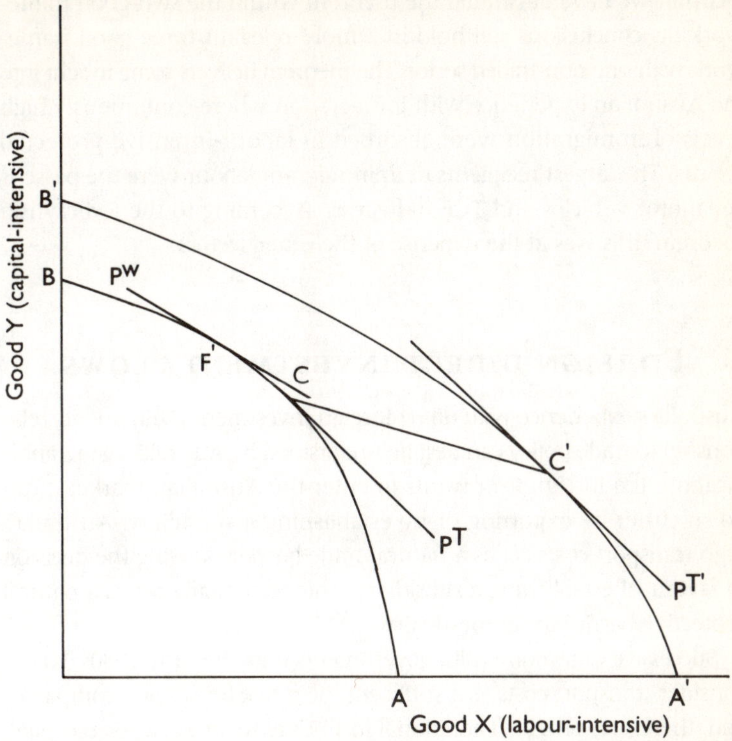

If we compare point C' to point C we obtain the Rybczynski theorem, which states that an increase in a factor of production will lead to an increase in the output of the good that uses the factor intensively while the output of the other good *falls* (see introduction and figure 0.4). At this point it will be useful to note that Rybcynski's theorem also holds for an exogenous increase in any factor of production. Suppose the economy was initially at point C and labour migrated without any inducement. In this case the production possibilities frontier would shift out and the new production point would still lie to the south-east of C.

The Rybczynski theorem has wide implications for the effect of increases in factors of production on the pattern of specialisation and trade. The Rybczynski theorem says that an influx of labour will shift production in a country towards labour-intensive sectors at the expense of less labour-intensive sectors. The importance of the Rybczynski theorem cannot be overstated in the Australian context. Despite the fact that we have illustrated the theorem within the two-good framework, its conclusions still hold in a more relevant three-good framework with one non-traded sector. The theorem delivers some insight into the Australian experience with immigration where continuously high levels of immigration were absorbed in labour-intensive protected sectors. The largest recipients of imimmigrant labour were the passenger motor vehicles and TCF industries. According to the Rybczynski theorem, this was at the expense of the other sectors.

FOREIGN DIRECT INVESTMENT FLOWS

Australia's experience with direct foreign investment (DFI) and its relationship to trade policy can best be understood by Australia's geographic location. If a foreign firm wants to enter the Australian market it can do so either by exporting or by establishing a subsidiary. Australia's high transport costs act as a natural trade barrier skewing the decision in favour of establishing a subsidiary. They essentially act as a natural protective barrier for domestic firms.

Successive Commonwealth governments until the early 1980s did not consider transport costs as a sufficient incentive for foreign companies and, therefore, provided additional incentives to attract foreign capital. Before World War II such incentives came in the form of protection instruments such as local content schemes or tariffs. Foreign companies were essentially coerced into establishing enterprises in Australia in order to be competitive in the domestic market. To some extent the case could be made that the line of causation should be reversed and that foreign investment was a *quid pro quo* for protection. There are very few studies to support this claim, but Brash (1966), in his famous study of American investment in Australia, surveyed a hundred American companies and asked them why they invested in Australia. Fifty firms admitted that tariff barriers were one of their top three reasons for establishing subsidiaries.[7]

It is also important to note that in the absence of transport costs and protection, Australia remained an attractive place to invest. Immediately after World War II Australia was growing rapidly and continued to grow until 1970.[8] The fact that companies could take advantage of a growing market turns out to be one of the most important determinants of investment. Brash's study confirms this point; eighty-nine firms in his survey indicated that market size was one of the top three reasons to invest (Brash 1966:36).

How different has Australia's experience with DFI been relative to other similar countries? According to data from the United Nations, Australia's experience with direct foreign investment does not seem too dissimilar from that of other countries. DFI stocks in Australia are currently 20 per cent of GDP. Canada has a slightly lower value, and the DFI/GDP ratio for the Netherlands is about the same as Australia's. Australia does differ from other countries with respect to the contribution of foreign capital to domestic capital formation. This ratio for Australia is 9.1 per cent. The UK, Benelux and Netherlands are the only developed countries with higher figures.[9]

Australia's traditional source of direct foreign capital before World War II had been the United Kingdom. Given its colonial background it was only natural to retain strong commercial links with the United Kingdom. After World War II there was a dramatic rise in the flow of capital from the United States. This was a global phenomenon, and Australia was one of many countries to receive a large injection of American capital.

Throughout the 1980s, however, the importance of both the US and the UK diminished as the importance of Japanese investment rose. In 1991 each of the three countries held about 20 per cent of Australia's total foreign investment. In 1981 Japan only accounted for 8.7 per cent while the US and the UK accounted for 27.6 per cent and 30.8 per cent respectively.

Similar figures can be obtained when data on changes in the level of direct investment in Australia are examined. In 1981 Japan accounted for only five per cent of total direct investment, yet by 1991 it accounted for 16.2 per cent. Comparable figures for the US and the UK are 35.5 per cent and 36.5 per cent in 1981, and these figures dropped to 26.9 and 25.2 per cent by 1991. In figure 6.2 we illustrate these data for 1981 and 1991. If this trend continues Japan could soon become the largest shareholder of foreign capital in Australia. It is also interesting to note that Australia's experience is not inconsistent with other countries. The boom in Japanese investment coincides with a global boom in DFI flows.[10]

Figure 6.2 Changes in the shares of foreign direct investment levels by country, Australia, 1981 and 1991

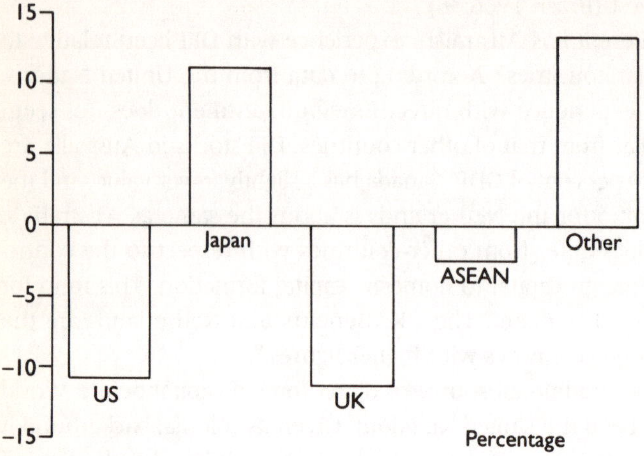

Source: Australian Bureau of Statistics (1981,1991), *International Investment Position Australia*, Cat. No. 5305.

Figure 6.3 shows the sectoral distribution of DFI. It is clear that the service sector has been growing at the fastest rate in the past decade (see chapter 5). The figure also indicates quite clearly that the demise of the manufacturing sector (see chapter 4) and the changing sectoral composition of Australia's GDP are affecting DFI flows. This observation, however, is not consistent with global events. Japan's increase in capital flows during the 1980s, especially to Europe and the US, did not exclude the manufacturing sector as it did in Australia. With most of the DFI in retail and finance, the figure could have been much higher if Japan had more confidence in the Australian manufacturing sector.

Japan is also not the only country in Asia that regards Australia as increasingly important. The ASEAN countries have also increased their foreign investment levels. Recent figures show that Singapore and Hong Kong are the two largest investors from this region; the bulk of their investment is concentrated in the finance and services sector.[11]

In addition to the shift in the source of foreign capital there has been a strong change in its sectoral composition. The manufacturing and mining sectors have consistently hosted the largest shares of foreign capital. In the past decade manufacturing's share has declined rapidly, and there has been a dramatic rise in the importance of the service sector. A recent study by the Bureau of Industry Economics reports that manufacturing held up to 77 per cent of DFI flows and services only 11 per cent in 1956–57. In figure 6.3 it is clear that these figures had been reversed by the early 1990s.

Perhaps the most important aspect of figure 6.3 is the noticeable decline in the role of the manufacturing industry as a host. This observation is not surprising in the sense that it supports the view that foreign capital was induced into the industry. Now that the process of trade liberalisation has begun manufacturing continues to decline and become less attractive to foreign investors. At this stage it is difficult to predict at what level the decline will stabilise. Declining protection levels in the automotive sector continue to put pressure on foreign investment in that industry. Ford Australia has announced that it will cease production of the Capri and will begin a program of plant shutdowns

Figure 6.3 Foreign direct investment by industry, Australia, 1991

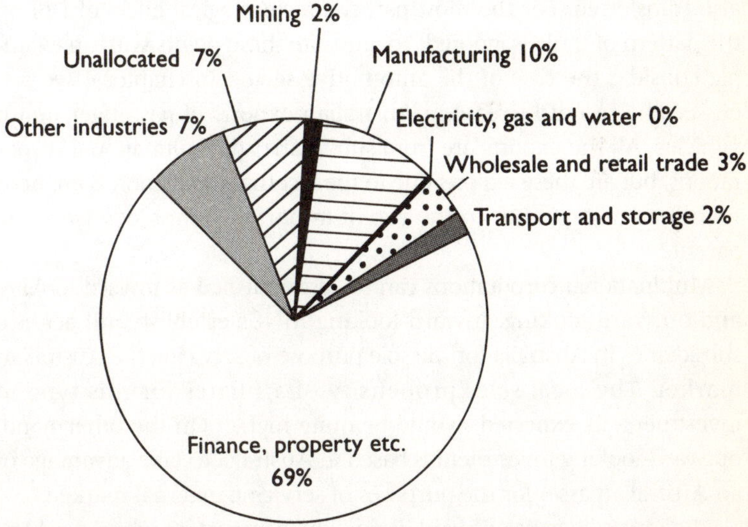

Source: Australian Bureau of Statistics (1991), *International Investment Position Australia*, Cat. No. 5305.

and retrenchments. It is important to note that these developments are related to Ford's lack of success in the export market. Their domestic market share remains strong, and there is little doubt that Ford intends to remain in Australia for the long run.

During the same week that Ford announced their plant closure Mitsubishi Australia announced that they were going to expand and modernise their Tonsley plant. Their success in the export market provides some empirical support to the Krugman (1984) proposition that import protection could lead to export promotion.

THE EFFECTS OF DIRECT FOREIGN INVESTMENT ON THE CURRENT ACCOUNT

Economists are still trying to understand the effects of foreign direct investment on the pattern and volume of trade. The principal problem in addressing these issues is that DFI is basically a 'bundled' factor of production. In addition to capital itself there are firm specific advantages in the form of technological knowledge or managerial expertise that are also transferred. For the most part the technological effect of DFI on the pattern of trade is ignored. To illustrate these points with an example consider the case of the automotive sector. In chapter 4 we discussed the recent increase in Australian exports of passenger motor vehicles. All the exports are from subsidiaries of multinational corporations, but are these exports due to the fact that production is financed with foreign capital or to the use of imported technology from the parent?

Multinational corporations can be distinguished as inward-looking and outward-looking. Inward-looking MNCs establish and acquire subsidiaries in Australia for the sole purpose of servicing the Australian market. The local sales propensity of affiliates for this type of investment, as expected, would be quite high.[12] On the other hand, outward-looking investment is based in Australia to take advantage of an Australian asset for the purposes of servicing external markets.

The inward versus outward distinction is important when conducting an analysis of the trade impact of foreign direct investment. Inward-looking DFI imports headquarter services and intermediate inputs, but does not export the final product, resulting in a negative impact on the balance of payments. Outward-looking DFI not only imports intermediate inputs but also exports the final product. In this case the net effect on the balance of payments would be ambiguous, but the volume of trade would be unambiguously higher.

Very little data are available on the operations of multinational affiliates in Australia. Bora (1992) uses data on the operations of US affiliates in Australia to show that the share of local sales in total sales is quite high; the figure is consistently 20 per cent higher than the average for developed countries. Drysdale (1993) develops a similar index for Japanese affiliates and finds that the figure is considerably lower. In the 1980s his figure did not rise above 55 per cent, whereas for the American affiliates the figure did not fall below 80 per cent. With such different data it is difficult to formulate a specific conclusion about the effects of multinationals on exports and imports, but it does highlight an excellent avenue for research.

Labour flows

The contribution of immigrants to the Australian economy has been the subject of one of the most controversial debates in Australia. Until recently, few people questioned Australia's policy of attracting immigrants but many had strong opinions about the composition of the immigrant intake.[13] In a survey of past opinion polls on the topic Goot (1988) reports that this selective attitude has been present since the first poll on the topic of immigration was taken in 1943. Despite the 'populate or perish' attitude of the government in the design of its economic policies, Australians have always been clear about their 'ideal' immigrant. The role of trade policy as part of Australia's policy mix was clearly to establish an excess demand for labour, and the government would then formulate criteria for the selection of immigrants to fill these vacant positions.

The Brigden Report was the first to espouse formally the income redistribution and wage effects of tariffs (Anderson 1987a:181).[14] An increase in protection would raise the demand for labour in the protected sector, which could be filled by increasing the intake of immigrants. The problem with this line of reasoning is that the sectors that required protection would be unable to compete in global markets. In the

Australian case, the level of skill expertise required to fill the vacant jobs was quite low. For the most part jobs in the labour-intensive unskilled sectors, such as textiles and clothing, footwear and passenger motor vehicles, were continually characterised by high turnover rates. Immigrants perceived these jobs as a stepping stone to higher-paying work elsewhere. In the presence of an immigration policy based on ethnicity and skin colour, highly skilled immigrants who did not meet these criteria were not allowed entry. Instead unskilled immigrants who met the criteria were allowed entry and found employment in the protected low-skill sectors.

Pope (1982) argues that in addition to the usual costs of protection (see the introduction) Australian society had to bear the cost of hosting immigrants with low skill levels. To establish this point he uses 1971 data to correlate effective rates of protection with the ratio of foreign-born workers decomposed into those from Britain and those from outside Britain. Pope's estimates yield a correlation coefficient of 0.72 for the first group and 0.80 for the second group. He concludes that this is evidence of the minimal contribution of immigrants because they are employed in low-skill and low-wage positions.

Interesting as Pope's results are, they do not indicate whether the immigrants in these sectors were underemployed. The tasks that they were performing required a low skill level, but they also could have been the only employment available. As soon as a better-paying task, more suited to the skills of the immigrant became available, they would leave the industry. The primary problem with an interpretation such as the one delivered by Pope (1982) and CEDA (1985) is that the imposition of protection should be addressed, not the selection of immigrants.[15] The positions were created by using tariffs and quotas, and they needed to be filled. Immigrants by definition arrive in Australia without employment and so they would be attracted to any vacant position.

From an international trade perspective, the skill level and participation rates of immigrants are important to the extent that they affect Australia's exports and imports. Exports are affected through the immigrants' enhancement or deterioration of Australia's comparative advantage. On the import side, it is their consumption pattern that will determine its volume and pattern.

Consider first the supply-side effects of immigrants. Regardless of whether immigrants were induced with protection or assisted with their passage, once they are in Australia they first need to be employed in order to contribute to the Australian economy. A study of immigrants by the Bureau of Labour Market Research (1986) concluded that until the 1980s immigrants had higher participation rates in the workforce, but it would take two or three years, depending on their gender, to become fully established. This means until that time immigrants were essentially consumers of public and private goods but not producers.

Once immigrants become employed there is the issue of whether they contribute more to production than an Australian-born worker. As could be expected, this issue has been debated extensively. Wooden (1994) provides a survey of the relevant studies, and interested readers would be wise to consult his paper. He concludes that, notwithstanding the measurement problems present in this type of analysis, the general conclusions of a study by Withers (1988) still holds; the skill level of the immigrant population exceeds that of the Australian population. One should also consider this result with caution since it is also sensitive to the composition of the intake. As Australia continues to allow entry to individuals under the family reunion and refugee program the average skill level of immigrants decreases.

On the import side there is the unambiguous claim that immigrants will increase imports because their marginal propensity to import would be greater than zero. What is important in the context of the effect on the current account is whether immigrants have a tendency to import more or less than Australian-born residents. In other words, we can expect an immediate increase in imports since there are more people living in Australia, but will there be an additional impact from a change in consumption patterns? Some attempts have been made to answer this question using computable general equilibrium techniques to estimate the effects of migration on the balance of payments. These are surveyed in Wooden (1994), and an intuitive explanation is also contained in Argy (1990). The general conclusion is that the combination of the import and export effects would increase net imports.

This conclusion is interesting, but it still ignores Australia's shifting comparative advantage and trading pattern. The first chapter showed how Australia has increased its economic linkages with Asia. Recent studies on the composition of Australia's immigrant intake also mirror the same pattern. In fact Australia's immigration pattern has historically been linked with its commercial ties. Does this mean increasing cultural linkages with Asia are necessary to continue and strengthen Australia's trade and investment links? Another important issue is the one we raised before about using trade policy for immigration purposes. Now consider the reverse and ask to what extent immigration policy can be used as trade policy.

The answers to these questions depend on the extent to which the lack of cultural linkages can be perceived as barriers to trade. For example, the sale of Australian goods to Japan will require the Australian company to have some knowledge of the Japanese language, distribution system, government taxation network and legal obstacles. The firm could hire and train an Australian, or it could hire a Japanese individual with the required knowledge. Immigration policy could then be used to help overcome trade barriers in Japan, and it would be an example of trade policy. We are not arguing for the use of immigration policy in such a manner, but highlighting the links between the two.

Today, in a general climate of unemployment and trade liberalisation, Australia's immigration policy continues to capture the attention of the public. Now, however, the issue is not the composition of the intake but whether there should be an intake at all. This argument is predicated on the thesis that immigrants would compete with resident Australians for jobs. In an economy characterised by a limited number of jobs, increasing the number of immigrants would only serve to exacerbate the unemployment situation. There is no clear solution, and the debates will continue. One important change in the debates, however, is the clear recognition that trade policy should not be used as an instrument of immigration. In the second section we discussed the importance of trade policy in the inter-war years to Australia's plan to populate the country. Today, the sentiment has changed, and those who advocate high protection levels are interested in retaining their current employment status, regardless of whether the competition is from imports or from immigrants.

POLICY IMPLICATIONS

Given the established relationship between protection and the migration of capital and labour through the Stolper–Samuelson theorem, to what extent has Australia's protection/immigration policy been successful? Surprisingly, despite the extensive literature on immigration, there are few studies addressing the issue. There has been some theoretical work that lays the foundations for its analysis. The most notable contributions are those of Johnson (1967) and Brecher and Diaz-Alejandro (1977). Both show that a small country that uses protection to induce factors to emigrate to their country might be economically worse off. The result might sound paradoxical in the sense that a country should become better off as it accumulates more factors of production, but the important point to note is that the presence of the tariff creates a distortion in itself. Therefore, as Johnson (1967) concludes, if the new immigrants or capital flow to the protected sector, they will increase the inefficiency in the sector with the potential of exhausting the gain from having the factors.

The intuition behind the Johnson and Brecher–Diaz-Alejandro results is illustrated in figure 6.4. Production before the increase in factors is given by point C and consumption is given by point N. An influx of labour induced by the tariff will cause the production possibilities frontier to shift out along the (dashed) Rybczynski line CR. We have drawn the line so

Figure 6.4 Effect of immigration in the presence of a tariff

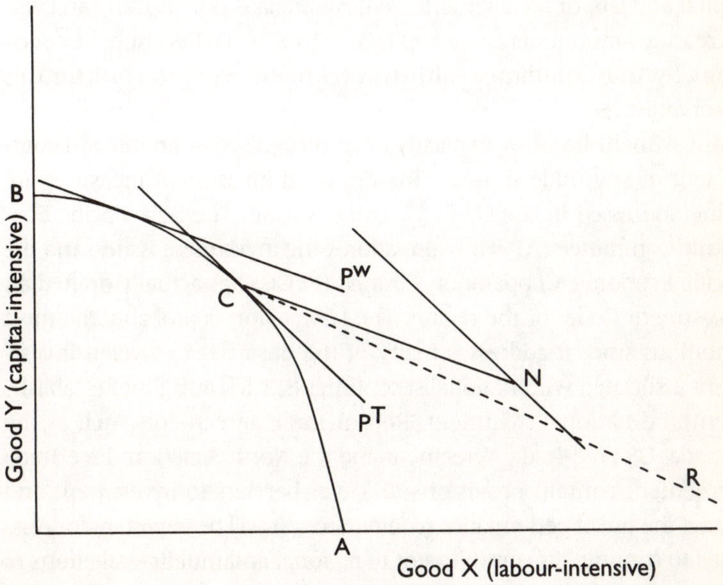

that it is steeper than the world price line so that small increases in the labour supply will result in the economy being worse off when welfare is evaluated at world prices. Of course, if the Rybczynski line was drawn flatter than the world price, line this result could not hold and the country would be better off.

It is also interesting to note that the conclusion cannot be extended to the case for any tariff-induced factor flow. If the level of migration is high enough the country could still become better off. Brecher and Diaz-Alejandro illustrate this point in a graph which shows that a country initially becomes worse off but, as factors continue to flow into the protected sector, welfare will increase and surpass the pre-factor inflow level of welfare.

CONCLUSIONS

An important element of this chapter has been the continued integration of Australia into Asia. In chapter 1 the importance of Australia's reorientation of trade links to Asia was highlighted, and the evidence that we have presented in this chapter reinforces the fact that trade, capital and labour are all interlinked. Australia is now hosting an ever-increasing amount of capital and labour from Asia. Recent policy decisions by the Commonwealth Government seem to confirm this observation.

Investment has now explicitly been recognised as an integral component of any trade strategy. Trade-related investment measures are being addressed in the GATT Uruguay Round. The Asia–Pacific Economic Committee (APEC) is also addressing investment issues, and the Pacific Economic Cooperation Council (PECC) has actually drafted an Investment Code for the region. The latter effort is probably the most ambitious since it addresses many of the obstacles to foreign investment associated with nationalist sentiments, such as right of establishment and national treatment. Recent trade agreements such as the Canada–US Free Trade Agreement and the North American Free Trade Agreement contain provisions to lower barriers to investment and restrict the use of performance requirements. It will be important for Australia to continue its commitment to regional and multilateral efforts to increase foreign flows of capital and technology.

Immigration issues will remain contentious. One important element of immigration policy in Australia today is that it does not discriminate on the basis of ethnic origin and colour. The 1988 Report of the Committee to Advise on Australia's Immigration policy distinguishes between three classes of immigrants: open, family and refugee. Such a succinct categorisation is useful when examining the impact of immigrants to Australia and consequently its trade performance and pattern. The latter two categories are outside the scope of this chapter, but the first category addresses directly the issues we have surveyed. If Australia wishes to continue on a path of growth, it must ensure that immigrants in the open category will contribute to areas in which Australia is experiencing a shortage of skills and complement those areas in which Australia is relatively skilled. This chapter has also highlighted the important complementarities between trade, industry and immigration policy.

FURTHER READING

On the role of factor movements and international trade theory the best reference is chapters 28 and 29 of Bhagwati and Srinivasan (1983). Factor movements are also treated in other trade theory books, such as chapter 7 in Krugman and Obstfeld (1994) and chapter 10 in Caves, Frankel and Jones (1990).

Lack of data has hampered much of the research on Australia's experience with foreign direct investment. The recent book entitled *Multinationals and Governments: Issues and Implications for Australia* by the Bureau of Industrial Economics provides an overview of Australia's recent experience with capital inflows. For a more detailed analysis of general trends in DFI flows the annual *World Investment Report* published by the United Nations is an excellent reference.

Australia's preoccupation with immigration issues has resulted in a wealth of literature. The second edition of *Australian Immigration: A Survey of the Issues* by Wooden et al. (1994) is an up-to-date and concise review of many of the relevant economic and social issues. Another recent publication that should be consulted for a broader perspective is Jupp and Kabala (1993).

7

TRADE AND MACROECONOMIC POLICIES

Colin Rogers

The process of the market liberalisation of the Australian economy in the 1980s began in 1983 with the floating of the Australian dollar and dismantling of exchange controls. It was followed shortly after by the deregulation of the financial system and the reduction of capital controls, which integrated the Australian financial and capital markets into the world market. However, the deregulation of money and capital markets sat uneasily with the still regulated goods and labour markets and exposed the need for structural adjustments in these markets also. The labour and goods markets were tackled under the Accord process and the general thrust to implement microeconomic reform. The need for microeconomic reforms (cost reductions) was exposed by the program of tariff reduction, which aims to have average effective rates of protection for the manufacturing sector down to 5 per cent by the year 2000. At the same time the Accord process has moved progressively to tie wage increases more closely to productivity and to allow some decentralisation of the wage-setting process in the form of enterprise bargaining.

Unfortunately, at a crucial point in the process of structural adjustment, macroeconomic instability increased significantly and threatened to undermine continued support for the process of market liberalisation.

Various factors contributed to the turbulence in the macroeconomic environment in the 1980s. In particular, Australia's inflation rate remained in excess of that of the OECD average; foreign debt increased dramatically after liberalisation of the money and capital markets; the terms of trade behaved like a roller coaster over the period 1985–87; the stock market crash of October 1987 and subsequent asset-price inflation, all had the potential to destabilise the macroeconomic environment. But the *coup de grâce* was delivered over the period 1988–90. In response to an asset price inflation, monetary policy was tightened significantly, and the economy slipped into recession in 1990 from which it only started making a hesitant recovery in 1994. The severity of the recession, together with the financial excesses and the social and economic costs it exposed, raised doubts in the minds of the public about the direction of economic policy in general and about the process of market liberalisation in particular. Not surprisingly, the significant increase in unemployment initiated pressure to halt or slow the process of structural adjustment in the goods and labour markets. With unemployment a major macroeconomic priority over the remainder of the 1990s there will be continuous pressure to slow tariff reductions to save jobs. Similarly, microeconomic reform, to the extent that it reveals overstaffing and inefficient work practices, will involve labour shedding and could also meet with resistance at a time when unemployment is unacceptably high.

In view of these developments this chapter re-examines the relationship between trade reform and macroeconomic policy from the perspective of the lessons that have been learnt from attempts at market liberalisation in other economies in the 1970s and 1980s. This experience has identified two important issues relevant to the Australian case: (1) it is now apparent that there is an important relationship between the macroeconomic environment and the success or failure of structural adjustment or liberalisation programs. Liberalisation programs attempted against a background of macroeconomic instability have been undermined by the need to restore macroeconomic stability; and (2) the sequence in which market liberalisation is attempted has important consequences in terms of the impact on real wages.

The first section deals with the relationship between trade liberalisation and macroeconomic stability. The second section outlines some of the arguments on the sequencing of trade and other market reforms, most notably capital market reforms. The general conclusion that emerges from these two sections is that macroeconomic stabilisation should precede any attempt at market liberalisation and that trade liberalisation should precede capital market liberalisation. The third section then considers the argument, due to Barry (1992), that Australia's liberalisation program in the 1980s got this precisely wrong! The capital market was liberalised before the current account (trade liberalisation), and both occurred before macroeconomic stabilisation, that is, before inflation had been reduced. Consequently, the welfare losses and other costs of Australia's experiment with market liberalisation appear, with the benefit of hindsight, to have been far greater than they need have been. The lessons from this experience for the 1990s are then briefly discussed.

TRADE LIBERALISATION AND MACROECONOMIC STABILITY

The relationship between macroeconomic policy and the liberalisation of trade has several dimensions. To begin with there is the question of the appropriate macroeconomic environment in which to undertake trade liberalisation.

A general, although not unanimous, principle that has been established here is that macroeconomic stabilisation should precede attempts at market liberalisation (Barry 1992:20; Villanueva & Mirakor 1990). Macroeconomic stability is a necessary prerequisite for financial market reform because it is now recognised that excessive interest rate movements contribute to financial sector fragility as a consequence of adverse incentive and selection effects (borrowers have an incentive to undertake riskier projects with a higher probability of default so the quality of loans deteriorates as interest rates rise). The consequences of the latter are compounded if prudential supervision is inadequate and/or incomplete, and the banking system faces a moral hazard problem if it misjudges the 'price' of insurance. Macroeconomic stability is therefore a prerequisite for financial market liberalisation, and difficulties can arise, as they did in Australia in the late 1980s, if these issues are not adequately addressed.

The experience with financial liberalisation in the 1980s demonstrates that, if the necessary condition of macroeconomic stability is not met, the costs of market liberalisation are likely to be higher than they would otherwise have been. There is always the risk that, if the costs become prohibitive, it is possible that the reforms will be suspended or abandoned, as was the case in several Latin American economies in the 1970s and 1980s.

Similarly, it is now recognised that macroeconomic instability can seriously impede attempts at trade liberalisation. If the macroeconomic environment is unbalanced, e.g., the rate of inflation is high and/or the government budget and the current account of the balance of payments are in chronic deficit, then this could well pose a threat to any program of trade liberalisation. For example, chronic inflation is likely to prompt a tightening of monetary policy at some point. Under a flexible exchange rate regime this could have a negative impact on attempts at trade liberalisation if it leads to a substantial appreciation of the real exchange rate. Similarly, the emergence of a chronic current account deficit and unemployment at the time of trade liberalisation will increase pressure for a reversal of the trade liberalisation policy. The gains from trade might be sacrificed as a consequence of the failure of stabilisation policy.

Instability in the financial sector has both direct and indirect consequences for trade liberalisation. Rising domestic interest rates and an appreciating exchange rate are the obvious direct effects but, in addition, the relationship between macroeconomic policy and trade liberalisation is further compounded by the question of the sequencing and timing of reforms. The sequencing of reforms can itself interact with stabilisation policy so as to undermine the reform attempt. To examine this issue in more detail the following section deals with the question of the sequencing of reforms and how the sequencing interacts with macroeconomic policy. The focus will be on the interaction between trade liberalisation and macroeconomic policy.

SEQUENCING OF REFORMS

Although the static gains from trade liberalisation are well understood, the question of the sequencing of reforms only received serious attention in the 1980s. Useful summaries are provided by Krueger (1986), Edwards (1986, 1989) and Michaely (1986). The question of sequencing arises when simultaneous liberalisation of all markets is not possible. This section focuses on the sequence of trade and capital market reforms.

From a macroeconomic perspective in an open economy the sequencing of trade and financial reforms has been treated as a question of whether the current or capital accounts should be liberalised first. Trade liberalisation clearly has a direct impact on the current account of the balance of payments, while liberalisation of financial markets and the removal of exchange controls have an impact on the capital account.

The question of sequencing raises a number of issues discussed by Krueger (1986), Edwards (1986) and Michaely (1986), but here we will concentrate on two to illustrate the significance of the issue. The first arises from the fact that each reform on its own will result in conflicting resource movements and consequently on income distribution when each reform is undertaken separately. For example, we might conjecture that trade liberalisation (removal of tariffs) will result in resources moving into the exportables sector. By contrast, liberalising the capital account when tariffs are still in place might result in resources moving out of that sector (see figure 7.3). If resource shifts involve some costs, this conjecture suggests that some attempt should be made to coordinate the two reforms so as to minimise the costs generated by resource churning. The second follows as a corollary from the Neary and Ruane (1988) demonstration that international factor mobility must raise the cost of protection. The corollary that follows from the Neary–Ruane conclusion is that capital account liberalisation which precedes current account liberalisation (tariff reduction) will impose a higher burden of adjustment on real wages than the converse sequence. Both of these propositions will now be examined.

Capital account liberalisation first (tariffs in place)

Following Edwards (1986) the consequences of capital market liberalisation can be analysed using the Melvin–Edgeworth–Bowley box. This framework ignores the inter-temporal aspects of capital market liberalisation, but it is nevertheless sufficient to illustrate the argument. The model consists of three commodities (exportables, importables and non-tradables) and two factors (capital and labour). The price of exportables is taken as the numeraire so $P_x = 1$. It is also assumed that at the time of liberalisation the domestic rate of time preference exceeds the world rate of interest. This means that a capital inflow results because domestic agents will borrow from abroad to increase domestic consumption. This simplifying assumption is made to avoid dealing with changes to the capital stock. Throughout, trade is not liberalised as importables are subject to a tariff, and the relative price of exportables and importables is constant, i.e. they may be aggregated into a single tradable good.

For the sake of simplicity it is also assumed that when the capital account is liberalised the capital inflow is steady and occurs for some time. Domestic expenditure exceeds income for some time, i.e. a current account deficit occurs and is the same in every period, and net foreign liabilities accumulate. As Edwards (1986:200) observes, this rules out the possibility of debt overshooting, which seems to be a common characteristic of experience with capital account liberalisation.

Short-run effects

A Salter-type diagram can be used to show that in the short run the increase in expenditure leads to an increase in the relative price of non-tradables to tradables, i.e. the real exchange rate (Pn/Pt) appreciates. In figure 7.1 exportables and importables are aggregated into tradables, and the short-run production possibilities frontier between tradables and non-tradables is represented by NT. When the capital market is opened up, under the current assumptions the consumption possibilities frontier becomes NT' and exceeds NT by the amount of the capital inflow measured in terms of tradables.

When the capital market is liberalised, and assuming that the income expenditure path corresponding to the original terms of trade $(Pn/Pt)^{\circ}$ is QS, consumption will move to S while domestic output remains at Q. Consequently an excess demand for non-tradables will emerge and the relative price of non-tradables will rise until a new equilibrium is attained at R (consumption) and P (production). The balance of payments is in equilibrium with a capital account surplus equal to the current account deficit represented by the distance RP when measured in terms of tradables. The rise in the relative price of non-tradables represents an appreciation in the real exchange rate.[1] The production of tradables contracts to OT_1 while the production of non-tradables expands to ON_1.

These results can now be illustrated in the three-good Melvin–Edgeworth–Bowley box diagram. Figure 7.2 shows the production functions in the exportables (lower left-hand origin), non-tradables

Figure 7.1 Short-run effect of a capital inflow

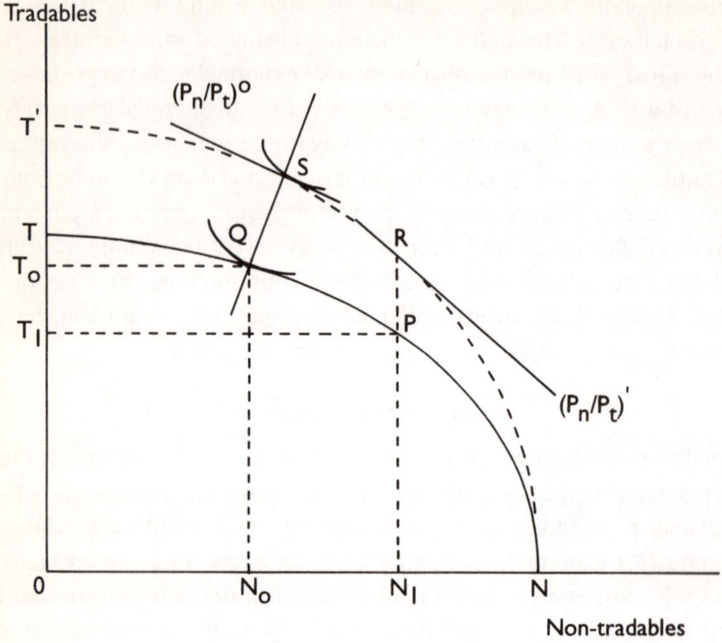

Note: A capital account surplus equal to the current account deficit is represented by the distance RP when measured in terms of tradables.

(top right-hand origin) and importables sectors. The isoquants for non-tradables are drawn relative to the origin O_n while those for exportables are drawn relative to O_x. The isoquants for importables are drawn relative to the origin at O_m. The initial equilibrium positions are given by A and G. This case also implicitly assumes that the capital labour ratios in exportables (x), importables (m) and non-tradables (n) are ranked in the following order: $(K/L)_x < (K/L)_n < (K/L)_m$. That is, exportables are the most labour-intensive sector while importables are the most capital-intensive.

The short run is characterised by the assumption that capital is specific to each sector. This assumption means that, although labour might be mobile in the short run, capital is not, so the allocation of capital to exportables production is O_xK_x, the allocation to the production of importables is K_xK_n and the allocation to non-tradables is 0_nK_n. This

Figure 7.2 The Melvin–Edgeworth–Bowley box

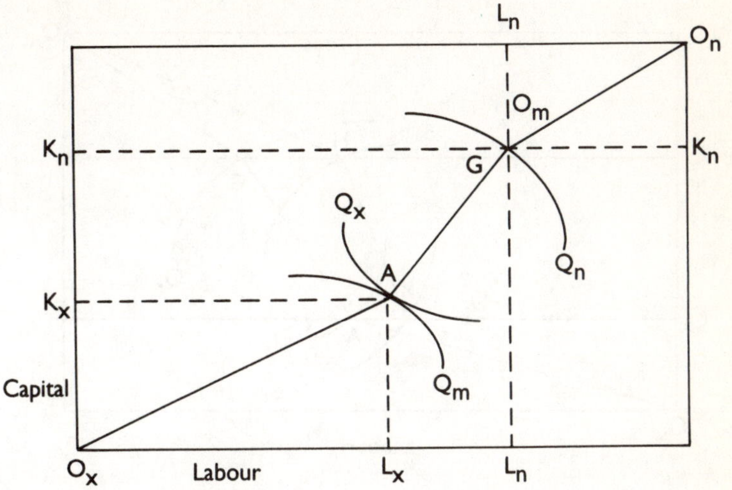

Note: The Melvin–Edgeworth–Bowley box represents the production functions for the three sectors: non-tradables, n, exportables, x, and importables, m. Output of exportables is indicated by point A on Q_x, output of non-tradables is indicated by point G on Q_n, and output of importables is indicated by point A on Q_m.

allocation cannot alter in the short run. Consequently, when the capital account is liberalised, expenditure exceeds income and the production of non-tradables expands while the production of tradables contracts. However, as capital cannot shift in the short run, the only way in which the production of non-tradables can increase is if production becomes more labour-intensive. Conversely, the production of tradables will become more capital-intensive. A possible short-run equilibrium outcome under the assumption of wage flexibility is illustrated in figure 7.3. The new capital–labour ratios are given by the dashed lines. The ray O_nH indicates that the non-tradables sector has become more labour-intensive while O_xB and BH indicate that exportables and importables production has become more capital-intensive.

The output of non-tradables increases from Q_n to Q'_n while that of exportables contracts from Q_x to Q'_x. The output of importables also

Figure 7.3 Short-run equilibrium after capital account liberalisation

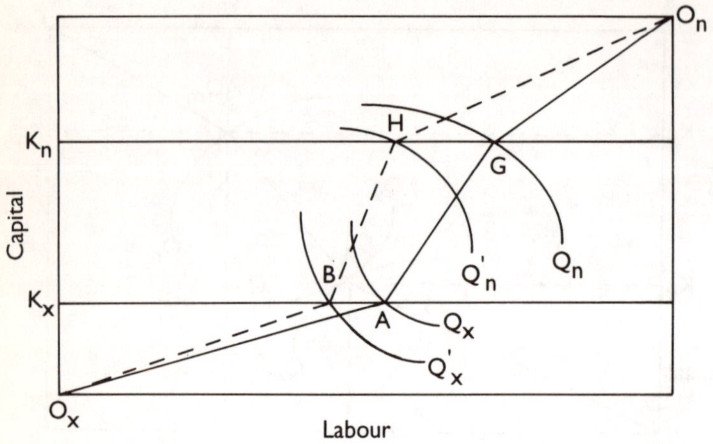

Note: Output of non-tradables expands to Q'_n while output of exportables contracts to Q'_x.

contracts but is not easily illustrated in figure 7.3. Note that in this setting the Stolper–Samuelson theorem does not hold in the short run.

Long-run effects

The long run is characterised in this model by the mobility of capital between sectors of the domestic economy, and relative prices of the three goods are completely determined by world prices, technology and the existing tariff. The opening of the capital account, therefore, has no impact on the relative prices of goods or factors. Nevertheless, as capital borrowed from abroad is used to finance higher consumption of non-tradables, the output of non-tradables will increase. Since full employment is assumed the increased output of non-tradables can only be achieved by a reallocation of resources from the tradables to the non-tradables sector. This means that production of exportables and importables will have to decline. The long-run effect of opening up the capital account is illustrated in figure 7.4.

Figure 7.4 Long-run effects of capital market liberalisation

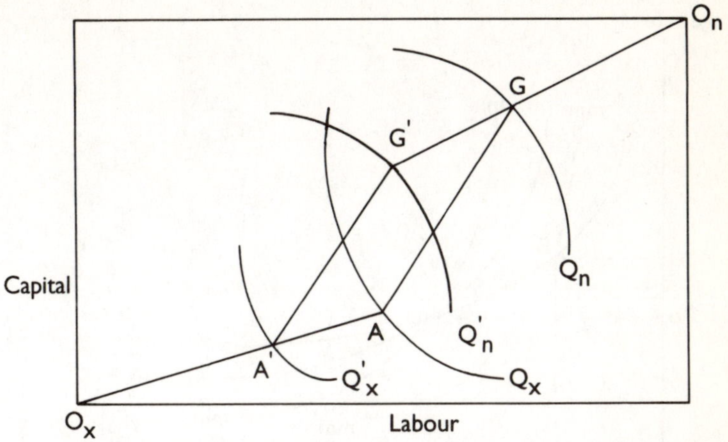

Note: Output of exportables contracts from Q_x to Q'_x. Output of non-tradables expands from Q_n to Q'_n. Prices of goods and factors remain unchanged.

Because there is no impact on the relative prices of factors the original factor proportions remain unchanged. The output of non-tradables expands from Q_n to Q'_n while the output of exportables contracts from Q_x to Q'_x. Capital and labour leave the tradables sector and are re-employed in the non-tradables sector.[2]

Current account liberalisation first (capital controls in place)

To demonstrate that tariff reduction generates outcomes the opposite of those under capital market liberalisation, we consider first the impact of a tariff reduction in the Lerner–Pearce diagram. Figure 7.5 illustrates the normal isocost curves for the three sectors (given the state of technology): non-tradables, nn, exportables, xx and importables, mm. The slopes of each curve reflects the capital–labour ratio in each sector in terms of the assumption made earlier that: $(K/L)_x < (K/L)_n < (K/L)_m$.[3]

Figure 7.5 Long-run effect of tariff reduction in the Lerner–Pearce diagram

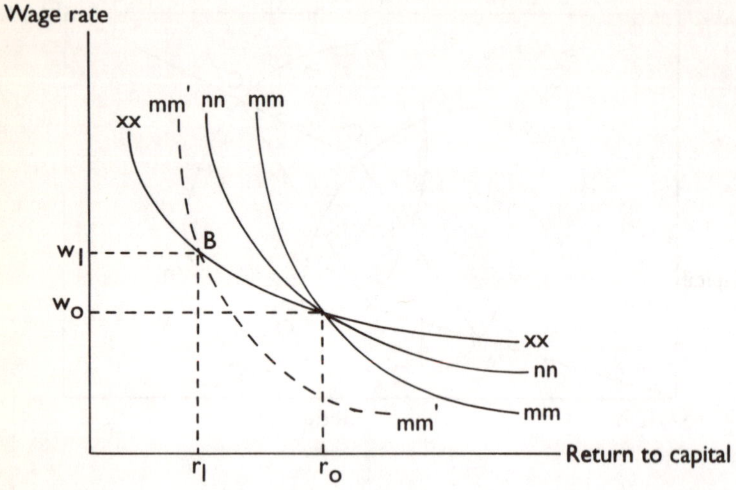

The reduction of tariffs on importables will shift the mm curve to the left because, to maintain equilibrium between domestic costs and world prices, lower combinations of returns to capital and wages will be required. The new long-run equilibrium wage and rate of return to capital are given by point B, which indicates that wages rise and returns to capital fall. Notice that this result is in accordance with the Stolper–Samuelson theorem, which suggests that reduction of a tariff in an economy where exportables are labour-intensive will result in higher wages and lower return to capital. Notice also that, as point B lies below the nn isocost curve, the price of non-tradables will have to fall.

Applying the results from the Lerner–Pearce analysis of the impact of a tariff reduction indicates that wages will rise and returns to capital will fall, which means in turn, under the capital intensity assumptions made, that all three sectors of the economy will become more capital-intensive. A possible long-run equilibrium solution after current account liberalisation is illustrated in figure 7.6.

Figure 7.6 Long-run equilibrium after current account liberalisation

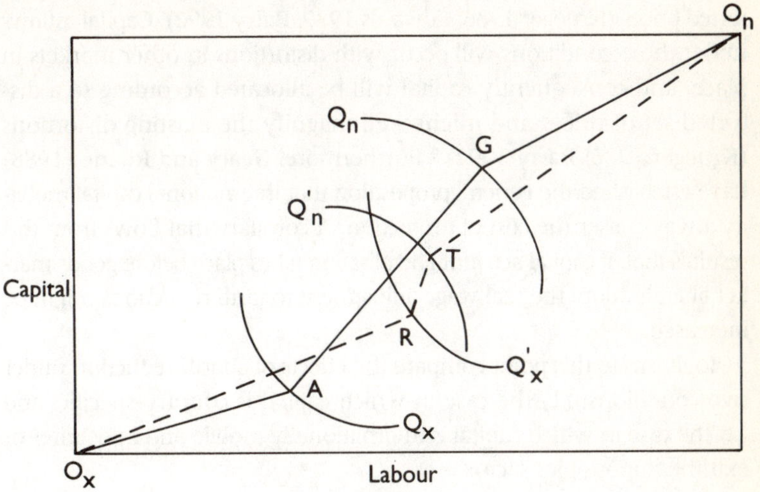

Note: Output of exportables expands from Q_x to Q'_x. Output of non-tradables expands from Q_n to Q'_n. Wages rise and capital rentals fall (measured in terms of all goods).

The long-run equilibrium illustrated in figure 7.6 indicates that the output of exportables has expanded (from Q_x to Q'_x), as has that of non-tradables (from Q_n to Q'_n), while the output of importables has contracted. The output of non-tradables increases because the relative price on non-tradables falls, and the positive income effect generated by trade liberalisation increases the demand for non-tradables. The capital intensity of all sectors has increased as can be confirmed by inspection of the rays O_xR, RT and O_nT.

The above analysis indicates that capital account and current account reform undertaken independently result in opposing effects on resource allocation in the exportables sector. Hence, to avoid unnecessary resource churning or, what is more likely, uncertainty about the consequences of market liberalisation, some attention should be paid to synchronising trade and capital account liberalisation.

The costs of incorrect sequencing

With tariffs still in place it has been generally recognised that capital market liberalisation means that capital inflows will be allocated at distorted prices (Krueger 1986; Edwards 1989; Barry 1992). Capital inflows under these conditions will occur with distortions in other markets in place, and consequently capital will be allocated according to a distorted set of prices and might well magnify the existing distortions (Krueger 1986; Barry 1992).[4] Furthermore, Neary and Ruane (1988) have established the general proposition that international capital mobility always raises the cost of protection. A corollary that flows from this result is that, if capital account liberalisation takes place before goods market liberalisation, the real wage adjustment to tariff reductions might be increased.

To illustrate this point compare the effects of a tariff reduction under two conditions: (1) the case in which capital is country-specific, and (2) the case in which capital is internationally mobile and may enter or exit the importables sector.

Applying the Neary and Ruane (1988) analysis, these two cases can be examined in a two-good model comprising exportables and importables. The impact of capital mobility is introduced by combining the traditional analysis of the labour markets in the two sectors with the unit cost curves for each sector (Lerner–Pearce diagrams). As before, the latter show the combinations of wage rate and return to capital consistent with normal profits in each sector. The given state of technology in each sector and the exogenously given output prices are the shift parameters on the isocost curves. Labour is treated as mobile between sectors, and capital is sector-specific but may be internationally mobile in the importables sector in this example. The wage rate must, therefore, be equal in each sector, but the return to capital might differ. The initial equilibrium in such an economy is illustrated for the real wage rate w_0 in figure 7.7.

1. *Capital controls in place.* Removal of a tariff on the importables sector shifts the demand for labour in the importables sector to the left from L_m to L_m', and the unit cost curve inwards to C_m'. To maintain labour market equilibrium the wage must fall to w' as the protected sector sheds labour that must be absorbed by the exportables sector. The wage rate w' then determines the return to capital in the importa-

Figure 7.7 Impact of capital mobility on the real wage effects of a tariff cut

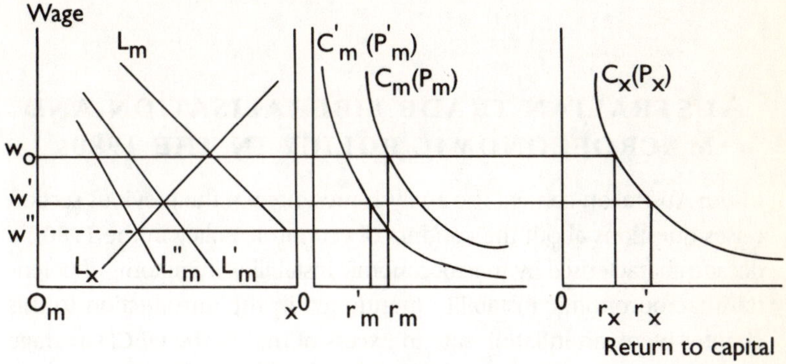

Note: Capital mobility increases the downward pressure on real wages when tariffs are reduced. With capital controls in place wages fall to w'. Under capital mobility wages need to fall to w".

bles and exportables sectors, which are lower, r'_m, and higher, r'_x, respectively, than before the tariff reduction.

2. *Capital account liberalisation.* With capital freely mobile between the protected sector and the rest of the world, the effect of a tariff reduction is to induce a capital outflow from the importables sector to maintain the return on capital in that sector at the international level, assumed, for the sake of simplicity, to be r_m. Since in this case r_m and C_m' are fixed, the demand for labour in the importables sector must contract even further and the real wage fall more than was the case with capital controls in place. This outcome is illustrated in figure 7.7 for the wage w" and demand for labour in the importables sector of L_m''. The impact on the importables sector of tariff reductions is far greater, in the sense that real wages must decline more in the case of capital mobility, than was the case with capital controls in place.

The object of this simple analysis is to demonstrate the point that the costs of adjustment might be increased if capital account liberalisation precedes current account liberalisation. The corollary is that any real wage rigidity will result in higher unemployment in the face of capital mobility.

AUSTRALIAN TRADE LIBERALISATION AND MACROECONOMIC POLICY IN THE 1980s

In the Australian context the analysis presented in the previous section raises questions about the conduct of economic policy in the 1980s; a decade characterised by macroeconomic instability. Symptoms of potential macroeconomic instability mentioned in the introduction to this chapter were: an inflation rate in excess of that of the OECD average and asset-price inflation in the late 1980s, volatility in the terms of trade, a significant increase in current account deficits and a quantum leap in foreign debt. All of these have the potential to increase the costs of any attempt at trade liberalisation in Australia. Relatively high inflation rates ultimately lead to a policy-induced monetary contraction, and chronic current account deficits inhibit the traditional demand management policy response to unemployment, which in turn increases the pressure to abandon trade reform.

As noted previously, the general conclusion from the literature on the stabilisation policy and the sequencing of reforms is that stabilisation should proceed first, with current account liberalisation followed by capital account liberalisation last. But as Barry (1992) argues, Australian economic policy proceeded in precisely the opposite sequence in the 1980s. The capital market was liberalised and the exchange rate floated before tariff reform was complete.[5] Finally, inflation was reduced to world levels in 1991, but only as a result of the excessive use of monetary policy aimed at the current account deficit and the growth of foreign debt!

On the face of it, the Australian experience with market liberalisation in general and trade liberalisation in particular would seem to be a textbook case of how not to proceed. As a consequence the costs of trade liberalisation have been higher than necessary and the resistance to further trade liberalisation has been increased. But, what is worse, the use of monetary policy to stabilise the foreign debt to GDP ratio not only increases the resistance to trade liberalisation, it also worsens the current account of the balance of payments.

The analysis of the negative impact of restrictive monetary policy on the current account has been well documented in the literature (Bewley & White 1990; Barry 1992; Rogers & Neal 1993). A monetary contraction pushes up domestic interest rates, attracting a capital inflow, which in turn results in an appreciation of the exchange rate. To the extent that the appreciation of the exchange rate changes the relative price of exportables and importables it worsens the structural net export position.[6] This conclusion also applies to a net debtor economy like Australia (most of whose debt is denominated in foreign currency), even if the valuation effect of the exchange rate appreciation on the net income deficit is taken into account. In other words, it is implausible (myopic!) to expect a monetary contraction to produce a reduction in the net income deficit sufficient to offset the deterioration in net exports (Barry 1992:14).

Lessons for the 1990s

In assessing the market liberalisation program of the 1980s in Australia it is difficult to avoid the conclusion that the costs have been higher than was anticipated, at least in part as a result of the type of problem discussed above. But these costs have largely been incurred and will be carried primarily by the long-term unemployed in the 1990s. The failures of economic policy also account, in the main, for the depth of the early 1990s recession and the weakness of the subsequent recovery. On current government projections the outlook for employment in the 1990s remains bleak; unemployment is not forecast to fall below 8 per cent until the end of the decade.

What can trade policy contribute by way of a solution to this problem? The simple answer would appear to be: not much in the short run. The contribution of trade policy is to deliver the gains from trade in the long run. In the short run trade policy will place additional pressure on wages and, in the face of real wage resistance, will be perceived as a cause of unemployment. Hence the outlook for trade policy in the 1990s is one of increased resistance to further reforms, irrespective of the success of the Uruguay Round. This resistance can be minimised, however, by increasing the flexibility of the labour market.

The success of trade policy in the 1990s is, therefore, intimately tied up with the success of the current push to labour market reform.

The message from the second and third sections of this chapter is that the liberalisation program in Australia has been badly implemented in the sense that the cost in terms of unemployment has been higher that it need have been. The message is reinforced by the argument presented by Forsyth (1993) that opening up the Australian economy will mean a decline in unskilled real wages. The point made by Forsyth involves a simple application of the factor price equalisation theorem of trade theory. Essentially what he suggests is that, as Asia expands its trade in largely labour-intensive manufactured goods, the world prices of these goods are falling. This means that, for all industrialised countries producing these goods, the price, wage and employment equilibria that existed before the expansion of Asian trade can no longer be sustained. From Australia's perspective the implications of the factor price equalisation theorem are that real wages in the old manufacturing industries will have to fall. As Forsyth (1993:3) puts it:

> All of this poses awkward dilemmas for the countries affected. They cannot continue to enjoy a combination of high employment, the established distribution of income, and the gains from trade ... If real wages are held constant, or increase, chronic unemployment amongst the low skilled will develop. If real wages for the unskilled are allowed to fall, the distribution of income will become more unequal.

As Forsyth also notes, there are no easy solutions to this dilemma. To the extent that there is a preference for an egalitarian outcome in Australia it can only be achieved at the cost of unemployment and/or some sacrifice of the gains from trade.

The pressures identified by Forsyth are directly related to trade policy, but they simply add to the risks already inherent in a situation in which trade liberalisation is implemented in an environment characterised by volatile and declining terms of trade for primary products against the background of a debt overhang. These elements have the making of a deadly cocktail if they provoke a stabilisation response to any deterioration in the current account. As has been the case in the past, macroeconomic stabilisation could well place trade liberalisation at risk in the 1990s. To minimise this risk labour market reforms, at the very least, are urgently required.

CONCLUSIONS

The experience of the 1980s has provided ample evidence that programs of market liberalisation are far more complex than was at first thought. On reflection the second-best perspective should have served as a warning that more caution was required. Nevertheless, the relationship between trade liberalisation and the macroeconomic environment is now better understood. In particular, macroeconomic stability is now recognised as a necessary prerequisite for a program of trade liberalisation, and there is some understanding of the consequences of particular reform sequences.

In the Australian context it can be argued that the pre-condition of macroeconomic stability was not met in the 1980s, and consequently the costs of trade liberalisation have been higher than was necessary. In addition, the sequence of reforms, particularly as they relate to the capital and current accounts, increased the costs of adjustment and placed additional downward pressure on real wages. It is also likely that liberalisation of the capital account before the current account increased resistance to further trade reform. Nevertheless, much of that cost has now been borne and is compounded with the cost of the recession of the early 1990s induced by the misguided attempt at using monetary policy to stabilise the current account.

Despite the fact that most costs have been borne, further trade reform in the 1990s remains at risk as a consequence of macroeconomic instability associated with the terms of trade. The major threat would appear to come from unemployment in the face of real wage rigidity in the labour market. The Neary–Ruane analysis clearly points to the risks of increased unemployment in the face of real wage rigidity if tariff reduction proceeds in the face of capital mobility. And, as Forsyth has pointed out, the difficulty is compounded by expanded trade with labour intensive economies (e.g. with South-east Asia). The same is true if the terms of trade for primary products remains under downward pressure in the 1990s. Both of these trends will place additional downward pressure on real wages (particularly unskilled wages) in the 1990s. Real wage resistance will compound the unemployment problem and might well induce increased resistance to further trade reforms.

Not surprisingly in terms of this analysis, part of the solution is to proceed with labour market reforms that will allow for relative and absolute real wage adjustments, which should minimise the unemployment consequences of the conditions likely to hold in rest of the 1990s. Wages policy and labour market reform will, therefore, play a central role as the liberalisation process initiated in the 1980s reaches its climax in the 1990s.

FURTHER READING

Rogers and Neal (1994) provide an introduction to open economy macroeconomics in the Australian context, and part 2 in particular deals with all of the reforms of the 1980s. Villanueva and Mirakhor (1990) give a useful overview of the lessons learnt from attempts at financial market liberalisation in developing countries together with a useful summary of the major theoretical issues. Edwards (1986) provides detailed discussion of the analytical tools employed in this chapter.

8

Trans-Tasman trade relations: Decline and rise

Ian McLean

The economies of Australia and New Zealand, with a combined population of 21 million and GDP of about $450 billion, are currently undergoing a process of integration impressive even compared with that occurring in western Europe. This process is centred on lowered barriers to commodity trade but also covers the progressive integration of labour and capital markets, liberalisation of trade in services, harmonisation of business regulations and the formation of a single aviation market. There has been no commitment, however, to the establishment of a common external tariff (customs union) nor to the creation of a currency union. The title of the 1982 treaty that underpins the process well describes what is involved: the Closer Economic Relations Agreement (CER).

The inclusion in this volume of a chapter on Australian trade with New Zealand is not based on the size of trans-Tasman trade flows. At no time has either country been the most important trading partner of the other. Primarily this is because throughout their history both have been commodity exporters with similar (geographical) patterns of trade, characterised in the nineteenth and first half of the twentieth century by a focus on Britain, subsequently on Japan and the United States. Rather, a study of trans-Tasman economic relations affords an opportunity to illustrate important features of Australian trade policy and a range of possible trade and beyond-trade arrangements in international economic relations. To this end, the chapter adopts a historical approach.

Between Australia and New Zealand there have occurred at one time or another conditions of unregulated trade in goods, services and factors; both narrowly prescribed and broad-ranging trade agreements; proposals for a customs union; and a type of currency union. In some periods, such as during the 1980s, economic relations have been harmonious, yielding benefits from closer economic integration. At other times, as in the 1920s, the state of trade relations was acrimonious, and the policies adopted at that time led to a reduction in the gains from trade.

THE SEVEN COLONIES OF AUSTRALASIA

In the initial phases of European settlement the scattered and small-scale regional economies of Australasia developed as components of an integrated trading system. Trade between the settlements on each side of the Tasman Sea was unrestricted, as was the movement of population, capital and enterprise. Symbolic of this degree of economic integration was the brief period during which the formal political administration of New Zealand was placed under New South Wales before New Zealand acquired the status of crown colony in 1841.

Early commercial relations

In 1793 two Maori were taken to New South Wales for the purpose of teaching the convicts how to dress flax. The project proved unsuccessful, and the Governor sent them back to New Zealand with gifts including axes, spades, seeds and pigs (Binney 1987:15–16). Labour migration (or guest workers?) and commodity trade (or aid?) across the Tasman had begun. Before the extension of formal British authority over New Zealand in 1840, private traders had operated between New Zealand and New South Wales or Tasmania, seeking timber and flax from the Maori, selling in return such items as muskets and blankets. Governor Macquarie is recorded as permitting duty-free entry to New South Wales of a cargo of timber and flax to aid the commercial activities of the missionaries among the Maori (Binney 1987:21). This decision was perhaps a harbinger of the liberal approach to trade policy between the two countries that was to characterise much of their subsequent history.

The closeness of economic relations between New Zealand and the Australian colonies after 1840 is readily illustrated. First, they were part of the British imperial system — political and economic. Before the granting of self-government during the 1850s there was free trade between the colonies — they were essentially regions of a greater British economy. There was no significant restriction on the movement of goods, capital or labour between them. Further, they were part of a wider imperial customs union and currency area in that they shared a common level of 'external' protection (from non-imperial imports) and that sterling (under the gold standard) was legal tender throughout.

Also binding the economies of the Australasian colonies was the similarity of their early development experiences. Wool was the first major export-oriented industry in New Zealand, and Australian pastoralists emigrated across the Tasman taking their merinos, capital and skills to underpin the development of the New Zealand pastoral sector through to the 1870s. Furthermore, gold figured prominently in the economies of both New Zealand and Victoria, with the favourable effects of the rushes spreading beyond their borders. The Victorian rushes in the 1850s stimulated agricultural development in New Zealand, while the discovery of gold in Otago in the early 1860s led to an outflow of labour from eastern Australia, at the same time as supplies for the diggings were sought in Melbourne, thus cushioning the Victorian economy during its post–gold-rush recession.

The close integration of the colonial economies at this time can also be observed in the capital market. The first permanent bank in New Zealand was the Union Bank of Australia, which opened a Wellington branch in 1840; the Bank of New South Wales (founded in 1817, now Westpac) crossed the Tasman in 1861, the same year the Bank of New Zealand was established partly in response to concerns at the influence of Australian banks. Indeed that influence seems to have been considerable: one estimate has nearly half the bank advances made in New Zealand in 1865 attributable to Australian-owned banks. Of course the principal source of investment capital for all the colonies was London, but there is evidence of on-lending or recycling of London funds through Australian banks to or from New Zealand according to economic conditions (Gardiner 1981:67–8).

Economic cooperation and Federation

A limited degree of independence in trade policy was made possible by the granting of self-government to the seven colonies of Australasia at various dates during the 1850s; limited because of the power retained by the Colonial Office in London to influence the external relations of each colony. Discussions between colonial governments on trade and related matters began in 1869 and covered preferential tariffs on intercolonial trade, ocean mail services and telegraph cables. The legislation granting responsible government restricted the ability of the Australian colonies to offer preferential tariff rates different from those of any other colony, whereas the New Zealanders were not so constrained. By 1873 the British legislated to permit the Australian colonies to fix preferential duties among themselves and with New Zealand. Suggestions from within Australia (which date at least from 1870) that the colonies negotiate a customs union were not, however, supported by New Zealand. The explanation for this reluctance appears to be that whereas intercolonial trade within Australia flourished, trade across the Tasman was of less significance; New Zealand's principal export market was already Britain (Dalziel 1987).

Broadly speaking, the Victorians evolved the greatest degree of protection in the regulation of their intercolonial and international trade during the late nineteenth century. New Zealand initially followed New South Wales in imposing only revenue tariffs. The situation changed in 1882 and 1888 when increases in the New Zealand tariff offered some protection to manufacturers. Although the rates were raised further and coverage widened in 1895 and 1907, the assessment has been made that by 1914 the general levels of between 15 and 25 per cent *ad valorem* were 'quite moderate' and that 'it is more than doubtful if these increases were sufficient to offset the fall in ocean freights that had taken place in the seventies and eighties' (Simkin 1951:51).

The most significant development of the period for Tasman economic relations was the move towards political federation among the seven colonies of Australasia. Under the constitution of the Commonwealth of Australia, the six colonies that finally chose to federate in 1901 accepted the abolition of all internal barriers to trade and gave the Federal Government the sole power to set the (common) external tariff. Federation thus created a customs union. Since the colonies were already part of the same (sterling) currency area, Federation did not immediately involve changes on that score, and there were no restrictions on the intercolonial movement of labour or capital.

New Zealand was involved in the initial discussions during the 1880s on proposals for Federation but sat on the sidelines during the 1890s. Despite a revival of public debate when Federation was imminent, New Zealand decided to remain apart. Analyses of that decision canvas several major reasons. Pressure from business groups in favour of joining was not strongly in evidence, presumably because exports to Australia constituted only about 12 per cent of total exports at the time compared to 83 per cent to Britain: 'For the majority of New Zealanders, federation with Australia did not make economic or political sense in the face of the expanding British market' after 1895 (Gardiner 1981:84). The decision might have turned more on the social and political differences emerging across the Tasman, such as the more generous treatment of the indigenous inhabitants, and on the more extensive body of labour market and social welfare legislation in New Zealand (Sinclair 1987b). It is important to recall, too, that whereas the 1890s was a period of economic hardship in eastern Australia, New Zealand had emerged from its 1880s recession, riding the prosperity that followed the upturn in world commodity prices after 1895.

AFTER FEDERATION: 1901–65

Following the New Zealanders' decision to stay out of the Commonwealth, trade relations across the Tasman acquired characteristics typical of international economic relations between any two states, including periods of considerable tension. The latter was most notable during the interwar years, described as a period of 'trade wars' between the two countries (Hawke 1987:115). Why had the trading relationship deteriorated? The direct impact of Federation on the New Zealand economy was not marked, although there was little growth in bilateral trade before the First World War. Attempts at this time to work out an exchange of preferential concessions were unsuccessful, the most likely reason being that the two economies were competitive rather than complementary in terms of what they traded. Hence, proposals from either party (New Zealand seems to have been the more frequent initiator) to liberalise entry into the market of the other for a particular commodity encountered resistance from well-established domestic producers threatened with import competition.

Australia raised its tariffs in 1921 and removed New Zealand from the British preferential rate previously applying. The New Zealanders retaliated, downgrading the preference they afforded Australian goods. This escalation at least produced negotiations leading to the first trade agreement between the two countries in 1922. However, its possible benefits were reduced by what Sinclair (1987c) has described as the 'Anzac plant war' during which attempts to gain improved access to the Australian market for New Zealand potatoes, fruit and hops were stymied by the use of non-tariff barriers — health and pest regulations. Negotiations in 1923 and 1924 to revise the 1922 agreement broke down. In 1927 trade tensions rose again as a result of a scheme to assist Australian dairy farmers by raising domestic prices. This required the exclusion of New Zealand produce, achieved by a lift in tariffs. Imports of New Zealand butter and cheese declined from $1.57m in 1927 to less than $10,000 two years later. The broader significance of these trade tensions should not, however, be exaggerated. In 1927, for example, the Australian market accounted for only six per cent of New Zealand's total exports, and New Zealand was the destination of less than three per cent of Australian exports (Sinclair 1987c:126).

Unsatisfactory rounds of bilateral negotiations over commodity trade were also a feature of the depression decade of the 1930s, but other developments probably were of greater significance to the Tasman trading relationship. In the period between 1929 and 1931 parity between the Australian and New Zealand currencies and sterling ended, with banks devaluing the former by more (30 per cent) than the latter (10 per cent). In 1933 New Zealand devalued again to 25 per cent below par (against sterling). As part of an effort to shore up their most important export market during a time of seriously diminished world trade opportunities, both Australia and New Zealand participated in the 1932 Ottawa Conference, and negotiated continued preferential access for their agricultural products to the British market at the expense of competitors from outside the British empire (such as Argentina). In 1934 the Reserve Bank of New Zealand was established, marking a step towards lessening the country's dependence on Australian-owned trading banks for the conduct of its international financial transactions.

New Zealand's response to the economic crisis of the depression included the introduction in the later 1930s of exchange controls and an extensive system of import licensing. Quantitative controls over imports were to remain the principal protective device until recent years; as discussed later, they are being phased out under CER. The broad context of their introduction in 1938 was the election three years earlier of a Labour government with a mandate for radical economic and social change, while the proximate cause was a foreign exchange crisis. Quantitative controls also served as an instrument of longer-term economic planning: 'What was a crisis measure was presented and defended as part of a planned insulation of the New Zealand economy from overseas influences, including enhanced protection for New Zealand industry' (Hawke 1987:122). The Labour Government wanted to accelerate manufacturing development thus providing a diversification of employment beyond farming and reducing dependence on imports for industrial supplies. New Zealand had held to a low tariff regime longer than Australia, possessed a much smaller domestic market and had no indigenous industrial enterprises comparable in scale to BHP, for example. Thus 1938 appears to be the point at which the New Zealanders adopted an inward-looking policy of industrial development, following the earlier Australian lead.

THE FREE TRADE AGREEMENT: 1965–82

After World War II trade relations between Australia and New Zealand were not at first a matter of major policy concern to either party, reflecting perhaps the bouyant international trading conditions facing commodity exporters. But during the 1960s the desirability of Australia and New Zealand improving their trading conditions was again discussed. This led in 1965 to the signing of the New Zealand–Australia Free Trade Agreement, which came into operation at the beginning of 1966. Although limited in scope, it marked a reversal of the interwar decline in economic relations.

The defining event of the 1960s for the long-term trade prospects and policies of both Australia and New Zealand was the move by Britain towards membership of the European Economic Community. The traditional concentration by Australasian agricultural exporters on the

British market was threatened by the protection afforded European farmers under the Common Agricultural Policy. The prospects for Australian dairy producers and for New Zealand dairy and meat producers were, transition arrangements aside, very serious. New Zealand was the more vulnerable due to the greater importance within total exports of those farm products most likely to face problems in retaining access to the British market and due to the greater reliance of New Zealand on Britain as an outlet for its products. Australia was less vulnerable to the economic consequences of the enlargement of the EC, partly because of the growth of mineral exports (virtually absent from New Zealand's post-war history), partly because of a more diversified portfolio of farm exports, and partly because the importance of Britain as an outlet for Australian exports was already in decline, offset by the rise in importance of the Japanese market, which followed the 1957 treaty on trade liberalisation between the two countries.

However, it should not be thought that New Zealand support for the Free Trade Agreement emerged as a response to prolonged recessed conditions in the New Zealand economy. A clear indication of serious problems in the economy only emerged in 1967–68 when the number of unemployed rose sharply and, for the first time since the depression of the 1930s, there was a net migration loss — primarily to Australia (Gould 1982:116–17).

The Free Trade Agreement

The aims of the New Zealand–Australia Free Trade Agreement of 1965 were expressed in the treaty as 'promoting a sustained and mutually beneficial expansion of trade' and ensuring 'as far as possible that trade within the Area [covered by the agreement] takes place under conditions of fair competition' (Australia 1966). The major characteristic of the agreement was, however, its restricted coverage. It did not take the form of a declared intention to move towards literal free trade by eliminating all barriers to the movement of goods after a defined transition period. Rather, freer trade across the Tasman was confined to goods enumerated in a schedule appended to the treaty (schedule A). The impact of the agreement on bilateral trade was, therefore, a function of the importance of these 'scheduled' items. Before assessing the impact of the agreement, other features of it should be noted.

Scheduled goods that faced import duties before the agreement would have these duties reduced to zero over a period of up to eight years, although provision was made for extending this transition period without limit (article 4). Where quantitative restrictions on imports of scheduled items existed — and, as noted earlier, this protective device was especially prevalent in the case of New Zealand — there was an injunction to reduce them 'at the earliest practicable date' (article 5). This objective, clearly reflecting New Zealand's concerns over the consequences of any rapid dismantling of its extensive import licensing system, was further qualified: progress in eliminating quantitative restrictions would be pursued 'to the extent permitted by the balance of payments'. The imposition of additional quantitative restrictions was prohibited (article 11), as were dumped or subsidised imports (article 10).

Other grounds for exemption from the terms of the agreement were also made explicit. One related to national development. It is relevant to recall that in the 1960s the view prevailing on both sides of the Tasman was that the expansion of manufacturing was vital to growth but unlikely to occur without a significant measure of import protection. Hence waivers could be applied to scheduled goods, or an item withdrawn from the schedule, 'for the purpose of establishing new industries or of encouraging the expansion of established industries' (article 8). Second, the temporary suspension of the treaty requirements was permitted where imports into a member state occurred 'in such increased quantities and under such conditions as to cause or threaten serious injury to its producers of like or directly competitive products', or even 'to prevent future injury' (article 9). These clauses were clearly capable of wide interpretation and reveal two governments not only concerned about the effects of freer trade on other national goals but also responsive to those firms and industries who might be exposed to increased competition.

Effects of the Free Trade Agreement

If the agreement was to provide for an expanding coverage of free trade across the Tasman, there had to occur an increase in the list of commodities entered on schedule A. However, the negotiation process established under the treaty made this difficult, giving maximum opportunity to affected parties to lobby against the inclusion of any item. In 1973 a system was arrived at of schedules of commodities *en route* to schedule A status, indicating the difficulty encountered in directly expanding the schedule A list. Despite successive meetings of officials, efforts to secure a significant widening of the application of the original agreement failed. It became evident that a different approach to trade liberalisation between the two countries was required.

Assessments of the sources of increased trade across the Tasman at this time concur that not very much credit can be given to the Free Trade Agreement. Bollard and McCormack (1985:18–19) note that, although New Zealand's share of exports to Australia rose during the seventeen years of the Free Trade Agreement, those of Australia to New Zealand peaked in the early 1970s. They also note that the share of trade attributable to items listed on schedule A was no higher in 1973–74 than in the years before the agreement. Accurately gauging the impact on trade of the modest liberalisation under the Free Trade Agreement is difficult in that other things pertinent to Tasman trade did not remain unchanged. There were movements in the exchange rates in the two countries; Australia undertook a general tariff reduction in 1973; New Zealand introduced export incentive schemes; and real unit labour costs did not move in parallel in the two economies. Lloyd's judgement is that 'during the life of [the agreement] movements in exchange rates have clearly had a much greater effect on competitiveness than the small average changes in levels of protection which [the agreement] achieved' (1984:8). He has also constructed estimates of the *real* exchange rate and found that, although wage rates rose faster in New Zealand than in Australia in the period 1966–82, the movement in the (nominal) exchange rate more than offset this, improving New Zealand's competitiveness in the Australasian market (p. 9). Thus it is difficult to conclude that the treaty by itself made much difference.

CLOSER ECONOMIC RELATIONS SINCE 1983

The background to the renewed political interest in trade promotion was the poorer economic performance recorded by both economies in the 1970s compared to the preceding two decades. By the late 1970s New Zealand's recurrent balance of payments crises had become especially severe, and a marked decline in relative living standards induced a sizeable migration of New Zealanders to Australia. To halt the economic slide and loss of human capital, the possibilities for further hitching

New Zealand's fortunes to the more prosperous Australian economy was one policy response. But by the early 1980s the limitations of the Free Trade Agreement as a framework for expanding trade across the Tasman were clearly evident.

The Closer Economic Relations Trade Agreement took effect at the beginning of 1983. In the period since there has occurred a sizeable increase in the volume of bilateral trade. There has also occurred a marked rise in investment flows across the Tasman, in migration, in the integration of capital markets and in progress towards the harmonisation of legislation affecting business and the formation of a single Australasian aviation market. By some measures, the Australian and New Zealand economies have moved further down the road towards an economic union than the European Community.

The CER treaty

The CER agreement was ambitious and broad-ranging, far removed in spirit from the limited and cautious Free Trade Agreement (Australia 1983). The introductory paragraphs spoke not only of 'the expansion of trade' but also 'the strengthening and fostering of links and cooperation in such fields as investment, marketing, movement of people, tourism and transport'. There was a 'commitment to an outward looking approach to trade', which, it was hoped, would lead to 'closer economic and trading links with other countries, particularly those of the South Pacific and South East Asia'. (As with the Free Trade Agreement, there was provision (article 24) to expand membership to additional countries.) The key objective (article 1c) was 'to eliminate barriers to trade between Australia and New Zealand in a gradual and progressive manner under an agreed timetable and with a minimum of disruption'. Whereas the Free Trade Agreement proceeded by listing those commodity groups to which the agreement to reduce barriers to trade was to apply, the CER began from the opposite position, announcing a move towards general free trade within a specified timetable, unless an exception was granted. That is, a comprehensive free trade area was planned, and any variation in the reduction in protection was permitted across commodity groups only with respect to the period over which free trade would be achieved. For items enjoying tariff protection, the CER required complete elimination within five years (article 4). The more complex task of liberalising items subject to quantitative restrictions resulted in a longer adjustment period: 'Quantitative import restrictions and tariff quotas originating in the territory of either Member State shall be eliminated by 30 June 1995' (article 5.14).

The commodities for which the move towards free trade across the Tasman would cause the most serious problems of adjustment and/or political sensitivity (the euphemism was 'special circumstances') were separately enumerated and permitted a 'modified application of the Agreement'. These commodities included wine, some timber products, carpet, iron and steel products, whitegoods, furniture, and motor vehicles and components, and the arrangements under which they were to be traded were separately negotiated. In addition, there was mention of the various agricultural assistance schemes in place in both countries, and wheat, fruit, sugar and dairy products were among the farm goods concerning which special provisions were written into the agreement. There was no requirement eventually to eliminate these schemes.

More decisive were the agreements to work towards the elimination of export subsidies and of preferences to local suppliers in the award of government purchasing contracts. The agreement to harmonise other trade-related issues ran to 'such matters as standards, technical specifications and testing procedures, domestic labelling and restrictive trade practices' (article 12) as well as to customs procedures (article 21). Moreover, the two governments agreed to cooperate 'to encourage or support the rationalisation of industries situated in the Area' (article 13). Industry advisory bodies in each country were enjoined to suggest the form any temporary assistance to an industry might take consistent with this longer-term objective.

The agreement was not subject to any time period nor did it require renegotiation or periodic reratification. Ministers of the two countries were to meet 'as appropriate', although there was to be a general review after five years of its operation (in 1988), regarding especially the progress in ensuring the removal by 1995 of all quantitative import restrictions. The participating governments clearly intended an irreversible process of integration to occur between the two economies.

Trade in services

In 1988 the two governments agreed to an extension of the CER to cover trade in services, and the protocol covering this extension came into force at the beginning of 1989. The broad objectives of the protocol included 'liberalising', 'expanding' and 'facilitating' trade in services, and the provision of a service was defined to cover 'production, distribution, marketing, sale and delivery'. Article 4 required each member state to grant to citizens of the other member state and services provided by them 'access rights in its market no less favourable' than those permitted in their domestic market. A number of provisions aimed to extend to services the liberalisation provisions previously negotiated for commodity trade, covering, for example, export subsidies and direct forms of government assistance (article 11). Further to illustrate the scope of the agreement, each member state 'shall encourage the recognition of the qualifications obtained in the other member state for the purpose of licensing and certification requirements for the provision of services' (article 9).

In a series of exemptions, each government listed areas to which the protocol would not apply. These included some aspects of the following: radio and TV broadcasting, postal services, coastal shipping and stevedoring, telecommunications, health and third-party insurance, and airport management and air traffic control. In some cases, such as coastal shipping and stevedoring, trade in services across the Tasman was too politically sensitive immediately to throw open to competition in the face of well-entrenched union or other monopolies. An important point, however, is that the exemptions had to be specified in the agreement; neither country could add new ones later.

Aviation

The CER agreement specifically referred to the 'strengthening and fostering of links and cooperation in such fields as ... movement of people, tourism and transport'. One area of substantial progress has been the formation of a single market for aviation services across Australasia. The negotiations have not formally been under the CER treaty but within its spirit. Discussions about freeing up the industry led in 1988 to agreement to expand the 'beyond rights' of the international carriers of each country on the trans-Pacific route. However, this agreement did not extend to any freeing up of domestic or trans-Tasman aviation, and Australia insisted on the exclusion of both domestic and international air services from the 1988 CER Trade-in-Services Protocol. A study of the effects of forming a single aviation market across the two countries reported in 1991 that

> under a wide range of scenarios, there would be sufficient welfare gains to Australia, New Zealand and, to a lesser extent, third countries if a single market were established. In general terms, consumers would benefit from reduced fares, improved service frequencies and greater choice of routes. Airlines would have new market opportunities, but some would lose profits due to increased competition. This could increase the pressure for the rationalisation of airlines or airline operation/capacity. (Bureau of Transport 1991:xi)

In 1992 Prime Minister Keating raised the issue of a single Australasian aviation market in his 'One Nation' statement. By August of that year a Memorandum of Understanding between the two governments set out a timetable for its realisation. The objective was to permit the domestic airlines of either country to operate within both and to link their services across the Tasman, a sector that would no longer be regarded as an international route. Although phased in, the process would be complete by November 1994. Thus Ansett may operate across the Tasman, integrating its operations in both countries. For its part, Air New Zealand may operate services between Australian cities in direct competition with Qantas and Ansett. Any new domestic carrier in either country would have similar rights. The determination of 'beyond rights' are still subject to negotiation, but partial liberalisation has already led to a decision by Air New Zealand to establish a hub at Brisbane to expand its service through Australia to Asian destinations.

Harmonisation of business regulation

Following the 1988 review of progress under CER, the harmonisation of regulations governing business activity in the two countries was begun. The relevant legislation is the Commerce Act in New Zealand and the Trade Practices Act in Australia, with the Commerce Commision and the Trade Practices Commission respectively being the agencies charged with their administration. For example, anti-dumping laws have been replaced with provisions relating to the 'misuse of market power' by sellers in either country, with cross-references to the relevant clauses in the other country's legislation. Further, the Federal Court of Australia can sit in New Zealand to consider claims of 'misuse of market power' against New Zealand sellers, with the capacity to issue and enforce subpoenas, injunctions and judgements. The New Zealand High Court may operate in Australia in reciprocal fashion (Baxt 1991:15).

Other agreements

Additional components of the Tasman economic relationship are progressively being brought into the CER process. One concerns government purchasing of goods or services. Historically, local suppliers have been given preference in the award of government contracts, although this discriminates against other suppliers and is inconsistent with the spirit of CER. In the 1983 treaty discrimination in the award of government contracts for goods was proscribed (article 11), but this applied only to the signatories, the New Zealand and Australian Governments. Following successful negotiations on the same issue between the Federal and State Governments of Australia, New Zealand was brought into the Government Procurement Agreement from June 1989, while from July 1991 the coverage was extended also to the purchasing of government services. Australian and New Zealand companies thus gained equal access to the market for supplying governments anywhere in Australasia.

A second example of the broadening of the CER relates to industry assistance. This was noted in the original treaty (article 13), but concerns by New Zealand manufacturers over the greater level of assistance to Australian industry led to an Agreed Minute on Industry Assistance being negotiated out of the 1988 review. Bounties paid on production for export within the area were abolished from July 1990, although the 'export facilitation' scheme applying to the Australian car industry was at first exempt. From the beginning of 1993, however, this scheme has not applied either to the export of motor vehicles from Australia to New Zealand or to components made there for the car industry.

As a result of a further review of CER in 1992, consideration was given to its extension to areas such as taxation and the mutual recognition of standards and occupational registration. Clearly the CER has become a process rather than a fixed outcome and has proceeded far beyond its initial focus on commodity trade.

Evaluation of CER

At the conclusion of their review of CER in October 1992 the Australian and New Zealand trade ministers claimed that 'CER is a resounding success' and 'has brought economic and social benefits and improved the living standards of both our countries. CER is an outward-looking agreement and an important part of our wider effort to encourage export awareness and make our economies more internationally competitive' (NZ Ministry of External Relations and Trade 1992). The second part of this assessment is more firmly based than the first.

Commodity trade across the Tasman expanded markedly in the years following the introduction of CER, but as a share of the trade of each country the impact has been larger, and with a more favourable impact on net exports, in the case of New Zealand. In 1983 New Zealand imports of Australian goods exceeded the value of exports to Australia by more than 40 per cent. This imbalance was eliminated over the next four years. From the Australian perspective, exports to New Zealand have grown only at the same rate as total exports — indeed they have been a remarkably steady share during the 1970s and 1980s at around five per cent. Even the exports of so-called elaborately transformed manufactured goods, for which New Zealand has been Australia's main market, have not grown any faster across the Tasman than to the rest of the world. In 1989 New Zealand took around a quarter of Australia's elaborately transformed manufactured exports, the same as in the early 1970s (Pinkstone 1992, tables 60 and 61). Of course, in the absence of CER the New Zealand market might have grown less rapidly, so a stability of export market shares is not by itself evidence that the treaty had no impact on commodity trade.

It is important to emphasise that CER covered all traded commodities not specifically exempt from treaty provisions and fixed deadlines for the phasing out of tariff protection (1988) and quantitative import restrictions (1995). These characteristics greatly reduced the ability of interest groups adversely effected by implementation of the agreement to stall or disrupt progress towards freer trade between the two countries, a situation that contrasts sharply with the arrangements under the earlier Free Trade Agreement.

Also, CER has not created a customs union: Australia and New Zealand continue to apply different tariffs on imports from third countries. And the lengthy phase-in period through 1995 on quantitative restrictions, together with scope for modified application of the agreement to specified commodities even after that date, indicates that neither is CER a free trade area strictly defined, although it could become one if negotiations succeed eventually in eliminating all exceptions to the general provisions of the treaty.

In other respects, however, the CER represents more than what is conventionally understood by a free trade area, especially in respect of the labour market. As previously described, there has long been a highly integrated labour market throughout Australasia. No visa is required to travel between the two countries; until 1981 no passport was required. Of greater economic significance is the unrestricted right citizens of one country have to work in the other — no work permits are required. The barriers to migration across the Tasman are thus relocation costs, just as they are for interstate migration of workers within Australia. Generally, there are few problems with acceptance of qualifications — educational, trade or professional. Indeed, some professional organisations are constituted on an Australasian basis. And students who cross the Tasman to attend university are charged the same fees as local students, not the higher fees applying to 'foreign' students.

In the capital market, too, there is a greater degree of integration than normally associated with a free trade area. We have already noted the long history of trans-Tasman involvement by Australian banks, and in recent years this has increased, with six of the top ten banks in New Zealand (including the two largest) being Australian-owned. This situation was highlighted in 1992 with the takeover by the National Bank of Australia of the Bank of New Zealand. But the integration of capital markets has not extended to a currency union or any other formal links between the two currencies, each of which continues to operate on a (slightly soiled) float.

Major (static) welfare benefits *for Australia* probably could not be expected to flow from CER. Theory would suggest that substantial gains from trade liberalisation would occur where two economies had relatively low volumes of bilateral trade because of high levels of protection but were *potentially* highly complementary in production. Another source of efficiency improvement arises where significant scale economies are unrealised because of the limited size of either domestic market but would be achieved in the integrated market. In the case of the Australasian market, neither of these effects is likely to be pervasive. The two economies are similar in economic structure and competitive in their commodity exports. Protection against the exports of the other country applied only to minor agricultural commodities (some dairy products). But there was protection of some industrial goods, and it is here that gains from freer trade are most in evidence.

Scale economies arising from the larger market have accrued primarily to New Zealand firms facing lowered barriers to selling in Australia, whereas the addition of the smaller New Zealand market makes less difference to the size of the 'domestic' market faced by most Australian producers. Of course, there are regional rather than national markets for many commodities even within Australia; and shipping costs on the Tasman remain a barrier to the closer integration of New Zealand and eastern Australian markets. However, if *dynamic* benefits are considered, there may be greater gains from freer trade for both countries as restructuring and relocation of production occurs in particular industries within the enlarged market. Recent investment patterns in the food processing, motor vehicle and whitegoods industries appear to reflect, at least in part, the new opportunities offered by the (re-) creation of an Australasian market.

An overall assessment of the growth or welfare effects of the moves towards integration of the two economies is difficult to arrive at because of the speed and extent of major structural changes occurring in both countries since the mid 1980s. This is especially true in New Zealand, where quite radical shifts in economic and social policy have occurred. These have included not only the floating of the currency and lowering of protection as in Australia but also privatisation or corporatisation of many public enterprises, a restructuring of the taxation system, a substantial dismantling of labour market regulations and a reduction in the comprehensiveness and generosity of the social welfare system. However, the faster pace of economic liberalisation in New Zealand, partly through CER, might have encouraged the process of economic reform in Australia. The New Zealand experience has featured prominently in Australian debates on the pace, nature and benefits of reform. More directly, competitive stimulus from across the Tasman is evident in several sectors of Australian business, such as in decisions on the location of new manufacturing plants.

FURTHER READING

The best general survey on relations between Australia and New Zealand, including economic relations, is to be found in the collection edited by Sinclair (1987a). Of particular relevance is the chapter by Lloyd (1987).

The Free Trade Agreement and CER agreement may be found in the Treaty Series published by the Australian Department of Foreign Affairs and Trade (see Australia 1966 and 1983). Over the last two decades the New Zealand Institute of Economic Research in Wellington has produced many studies on a range of topics related to the economic relationship with Australia, two of which are Bollard and McCormack (1985) and Bollard and Thompson (1987).

9

THE INTERNATIONAL POLICY ENVIRONMENT

Richard Pomfret

The preceding chapters have focused on Australia's trade policies and domestic determinants of trade. These are the most important but not the only influences on Australian trade patterns and performance. As chapters 1 and 2 show, Australian trade can be much influenced by the state of the world economy; it suffered from the global depression of the 1930s and benefited from the long boom of the 1950s and 1960s, as well as from the increases in primary product prices in the early 1950s and 1970s.

Through the colonial period and the early post-Federation years, Australian policymakers were keenly aware of the importance of access to export markets, especially that of Britain. As the twentieth century progressed, however, this concern became less dominant, and Australian policymakers seemed to respond only to domestic considerations. Even when imperial preferences were given a boost in the early 1930s it was in the Ottawa Agreement, not a Canberra agreement.

The most striking illustration of this insularity in trade policymaking was Australia's effective non-participation in the establishment of an open world trading system in the quarter century after the signing of the General Agreement on Tariffs and Trade (GATT) in 1947. During this period Australia, alone apart from New Zealand, distanced herself from the trade liberalisation undertaken by the other high-income countries.[1] Rather incongruously identifying with Third World primary product exporters, Australia demanded separate treatment, refusing to reduce her trade barriers protecting the manufacturing sector on the grounds that other GATT members were excluding agricultural trade barriers from the negotiations. The intellectual basis for this approach to trade policy collapsed in the 1970s when it became apparent that the most successful Third World countries were those that had unilaterally liberalised their foreign trade, while those clinging to an import-substituting industrialisation strategy were experiencing declining growth rates. As we have seen, Australian policymakers saw her past history mirrored in the latter group's experience and dramatically shifted direction on trade policy after 1973. The international counterpart to this policy change was a renewed activism within GATT, leading to a well-defined leadership role for Australia in the second half of the 1980s. The first section of this chapter deals with Australia's position in GATT, emphasising its position in the Cairns Group during the Uruguay Round of trade negotiations.

Just as Australia was embracing the GATT principles of liberal multilateral trade, the system was coming under threat, most notably from the actions of the USA, which had been its staunchest defender in the 1950s and 1960s. The threat came in part from a growing tendency of the major trading nations to sidestep GATT principles by unilateral actions such as anti-dumping or countervailing duties, which did not break the letter of GATT agreements, or by voluntary export restraint agreements, which were probably not GATT-compatible but which bought off potential protests (see the introduction). The threat to GATT also came from pressures in the USA and the EC to manage trade by making bilateral agreements with Japan and other Asian countries for market access (notably the US–Japan semiconductor agreement and a US–Korea agreement on insurance). A further trend towards bilateralism was seen in the proliferation and strengthening of preferential trading arrangements, especially the US–Canada (and then North American) Free Trade Area and the '1992' program to complete the EC's internal market.

Australia's reaction to these developments was first and foremost to emphasise its commitment to GATT, which provides the best protection of the rights of small trading nations. Australia also worried about being left out in the cold if the North American Free Trade Area (NAFTA) spread to South America, the EC spread to northern and eastern Europe, and Asian countries responded by forming a regional trading arrangement; in a world of three trading blocs, Australia would be one of the few kids without a bloc. The issues associated with regional trading arrangements and Australia's efforts to ensure that Asia at least remains committed to open multilateral trading are dealt with in the second section.

AUSTRALIA IN GATT

The post-1945 international economic system was heavily influenced by the United States, which emerged from the Second World War as by far the dominant economic and military power. US policy was driven by the view (held particularly strongly by Secretary of State Cordell Hull) that rising protectionism had contributed to the depth of the 1930s depression and had been a cause of the world war. The USA also held an ideological opposition to colonialism, which was expressed as a desire to eliminate imperial preference schemes. Thus, the United States exercised strong leadership to establish an international system based on non-discrimination and lower trade barriers. These principles were embodied in the GATT and in the articles of the International Monetary Fund, which sought to prevent the use of foreign exchange restrictions as a barrier to international trade.

The cornerstone of the GATT, stated as its first article, is that every signatory must accord to every other signatory all the trade privileges extended to its most favoured trading partner. The unconditional most favoured nation (MFN) principle simply states that countries should not discriminate between trading partners. There are some exceptions to this principle, notably article XXIV on customs unions and free trade areas, to which we will return in the next section, but the GATT clearly envisaged these exceptions to be of declining significance. The second guiding principle was to encourage transparency in trade restrictions and specifically to discourage the use of non-tariff barriers. There is a strong theoretical case for the position that if a trade barrier is to be introduced then a tariff is likely to be the least harmful instrument (see the introduction). Thus, the GATT aimed to establish the rule that non-discriminatory tariffs would be the only generally acceptable trade barriers, and if signatories broke the rules then other signatories were allowed to retaliate. The threat of approved retaliation is the only GATT enforcement mechanism, but it has been successful in discouraging both explicitly GATT-incompatible actions and unregulated trade wars such as occurred in the 1930s.

The initial impetus for many countries to sign the GATT was the promise by the USA to participate in rounds of multilateral trade negotiations to reduce tariffs. Specifically, the USA promised to reduce its own tariffs in the first rounds without requiring comparable actions from other signatories. Thus, in the late 1940s and 1950s US tariffs were drastically reduced from their high level of 1930. A major step forward came in the Kennedy Round of the mid 1960s when the European Community participated as a single negotiator and the USA, EC and other GATT signatories agreed to across-the-board tariff reductions. The Kennedy Round brought average US and EC manufacturing tariffs below 10 per cent, and in other major trading nations the average was only slightly higher (e.g. 12 per cent in Japan and 14 per cent in Canada; see table 9.1). The tariff reductions were taken even further in the Tokyo Round (1973–9), which was more drawn out than previous rounds because it covered non-tariff barriers for the first time (table 9.2).

Australia, as a GATT signatory, benefited from the improved market access that she received automatically under the MFN principle.

Table 9.1 Average manufacturing tariff, selected industrial countries, 1925 and 1970

	1925	*1970*
Australia	27%	23%
Canada	23%	14%
USA	37%	9%
EC	na	8%
Japan	13%	12%

Note: EC members' tariffs ranged widely in the 1920s from close to zero in the UK and the Netherlands to above the US average in Spain. Thus Australia was around the median of the countries in this table in 1925 but an outlier in 1970.

Source: Anderson (1987a).

Table 9.2 GATT rounds of multilateral trade negotiations, 1947–94

Date	*Round*
1947	Geneva Round
1949	Annecy Round
1950–51	Torquay Round
1956	Geneva Round
1960–62	Dillon Round
1964–67	Kennedy Round
1973–79	Tokyo Round
1986–94	Uruguay Round

Note: An international congress in Havana in 1948 agreed on a charter for an International Trade Organisation (for which the GATT had originally been intended as a draft), but in the absence of ratification by the US Congress the ITO fell into abeyance. Although the pre-Dillon rounds were multilateral, the European countries (like Australia) retained exchange controls throughout the 1950s, so their tariff reductions were nugatory.

Australia's role in the multilateral trade negotiations was almost entirely passive. She identified with the developing countries, which refused to reduce their own trade barriers on the grounds that they were necessary to develop a more diversified economy. More specifically, Australia used the lack of other countries' trade liberalisation in agriculture to justify her own retention of high trade barriers in manufacturing. Thus, while the manufacturing tariffs of other high-income countries were tumbling, Australia's remained roughly at the levels of the 1920s (averaging 23 per cent in 1970). Indeed, with the adoption of new non-tariff barriers in the 1960s (such as those described in chapter 4 for automobiles), Australia was not just passive but actually swimming against the trade liberalisation tide.

Australia's only active trade liberalisation during the 1960s (after the removal of quotas in 1960, and apart from the limited regional scheme described in chapter 8) was to apply for a GATT waiver to give preferential tariff treatment to imports from developing countries. This 1965 initiative made Australia a pacesetter in the area as the GATT allowed its members to introduce similar schemes under a generalised system of tariff preferences for developing countries (GSP) in 1971. It also reinforced Australia's role as champion of developing countries' interests in that preferential market access was a major demand of developing countries during the 1960s. In practice, however, GSP turned into a dead-end as a stimulus for trade-led economic development. The importing countries, including Australia, tailored their schemes so that preferential margins were high on items in which developing countries had no comparative advantage, but market access remained restricted on the goods that developing countries could produce competitively, notably textiles and clothing. The flaw lay in relying on unilateral action by the high-income countries, in which domestic interest groups ensured that preferential tariff reductions would do minimal damage to import-competing activities. Developing countries ended up obtaining little benefit from GSP and losing substantially by their absence from the GATT bargaining table. Only in the post-1986 negotiations did the developing countries participate actively in a GATT round, and by doing so they were able to add items of special interest to them (like textiles and clothing) to the agenda.

Australia learned a similar lesson. Absence from GATT negotiations allowed Australia to retain her protectionist trade policies, but these policies only resulted in self-inflicted wounds. The post-1973 unilateral trade reductions allowed Australia to show a reformed face with respect to trade liberalisation. In the 1980s Australia used its new status to bring agriculture back on to the GATT agenda.[2]

The basic problem with respect to agriculture and GATT was that the USA strove to maintain liberal principles for world agriculture trade in the 1950s and 1960s, but responded to domestic farm groups by requesting a waiver permitting use of GATT-incompatible import quotas on some items. The lesson here is that important trading nations create exceptions to the theoretically well-based GATT principles at their own risk. The US action provided a precedent for taking agriculture out of GATT negotiations, which was seized on by major importing nations; when the European Community drew up its Common Agricultural Policy during the 1960s little effort was made to ensure GATT compatibility, and when Japan undertook its general liberalisation of foreign trade controls in 1961 the Basic Agricultural Law allowed for agriculture to be an exception. The global food crisis of 1972–4, when several major producing areas experienced poor harvests and world prices reached peaks, reinforced the determination of the EC and Japan to retain high levels of self-sufficiency in food.

These policies were important for food exporting nations not only because potentially large import markets remained closed but also because the EC's price support system began to generate large surpluses (butter mountains and wine lakes). Disposal of these surpluses at cut-rate prices alleviated the EC's budgetary problems but also reduced world prices. Under pressure from its farmers the USA began to respond aggressively to the EC's surplus disposal by making subsidised wheat sales to countries in which the EC was disposing of its own surpluses (notably to Egypt in January 1983 and to the USSR in 1986). The subsidy wars further undermined world prices, especially for wheat, animal feedstuffs, meat and rice.

Australia's reaction to deteriorating world agricultural trade conditions in the early 1980s was abrasive. Deputy Prime Minister Doug Anthony stormed out of the 1982 GATT ministerial meeting, calling it a fiasco, after his failure to get agriculture on the agenda; other delegates viewed his petulant action with little respect (Higgott & Cooper 1990:609). This unsuccessful approach was replaced in the mid 1980s by a step-by-step approach avoiding animosity and recrimination. In August 1986 Australia organised a meeting of selected agricultural exporters at Cairns, and in September 1986 the Cairns Group was successful in placing agriculture on the agenda for the Uruguay Round of trade negotiations.

The Cairns Group brought agriculture back into GATT discussions. In large part this success was due to Australia's intellectual and entrepreneurial leadership. The fourteen members (Argentina, Australia, Brazil, Canada, Chile, Colombia, Fiji, Hungary, Indonesia, Malaysia, New Zealand, the Philippines, Thailand and Uruguay) are a motley crew with differing perspectives, and when it comes to hard bargaining they have different interests in specific items. The Latin American countries are especially desperate for export earnings to service their foreign debts, but some are also concerned about maintaining Third World solidarity with developing countries that import food. Canada, which initially shared group leadership with Australia, became distracted by its bilateral trade negotiations with the USA (see next section) and at times seemed more concerned about retaining its own high barriers to trade in dairy products, wine and other agricultural goods than opening export markets for its wheat. Australia has earned considerable credit for its diplomacy in keeping the group together and focused on the relevant issues of regulating (and eventually outlawing) export subsidies and improving market access.

Australia's own trade liberalisation and the country's leadership role in the Uruguay Round have set the scene for a new activism in international commercial diplomacy. The promise for the future is that Australia can take part in the many overlapping coalitions on trade policies to have a say in the future evolution of world trade law. Australia will, of course, never have the weight of the USA, the EC or Japan in international economic forums but, as economic power becomes more diffused, decisions will depend on forming coalitions, and middle powers will be difficult to ignore.

REGIONAL TRADING ARRANGEMENTS

An important exception to the MFN principle is that GATT members are allowed to form free trade areas or customs unions. The welfare analysis of such regional trading arrangements (RTAs) produces ambiguous conclusions because discriminatory reductions in trade barriers lead to both trade creation and trade diversion (see the introduction). From a global perspective non-discriminatory reductions in trade barriers are generally preferable because they reap the trade creation benefits while avoiding the misallocation of resources associated with trade diversion. Nevertheless, supporters of RTAs argue that they might sometimes be the vehicle for greater trade liberalisation among like-minded nations than is feasible in the slow-moving multilateral trade negotiations. This happened among the European countries that formed the European Communities (EC) in 1957 and is being used to justify the US–Canada free trade agreement, and its geographical extension to include Mexico (now known as the North American Free Trade Agreement, or NAFTA). In particular during the early 1990s RTAs were canvassed as a practical route to reducing non-tariff barriers, which were proving stubborn to agree on in the multinational trade negotiations under GATT.[3]

Non-members are likely to suffer from the formation of RTAs because the trade diversion involves loss of export earnings by non-members while any benefit from the increased income and demand for imports by the RTA members is of second-order magnitude. For Australia the economic effects of the EC's formation were minor. The six original members were not important trading partners for Australia so both the negative trade diversion effect and the indirect benefit of higher EC incomes were small. Much more important for Australia was the United Kingdom's accession to the EC in 1973, from which the trade diversion was significant, especially hurting Australia's agricultural exports. Nevertheless, the EC's formation and enlargement encountered little international opposition, largely because they have been part of a political process of European integration rather than involving opportunistic use of discriminatory trade policies to benefit members at the expense of outsiders. Some EC policies have come under fire, especially the Common Agricultural Policy, but the criticism is over the instruments used and the level of protection rather than the customs union.

The many other RTAs formed during the 1950s, 1960s and 1970s were of minor global relevance. A critical reason for this insignificance is that they mostly aimed to create a larger internal market for the members' protected industries. This motive led to unsuccessful RTAs because each member wanted its industries' exports to the partners to expand but did not want to buy overpriced or poor-quality goods from its partners. Thus, RTAs within Latin America, East Africa, South-east Asia and between Australia and New Zealand, which involved similar rather than complementary economies and hence were more likely to produce trade diversion than trade creation, all decayed.

The RTA issue came to prominence during the 1980s, partly because of the publicity given to the EC's program for completing the internal common market by 1992. Non-members feared that removing non-tariff barriers to trade within the EC would increase the discrimination against outsiders and might even be used surreptitiously to increase the EC's external trade barriers. Even more important for the changing perspective was a shift in US policy. Until the early 1980s the USA had been the staunchest defender of the MFN principle, but the Reagan administration began to adopt RTAs as a tool of commercial policy. The initial steps, a 1983 system of one-way preferences for certain Caribbean countries and the 1984 US–Israel free trade area, affected small amounts of trade, but the US–Canada free trade agreement, which came into effect in 1989, covered the world's largest bilateral trade flow.

When the USA and Canada began negotiating with Mexico to form NAFTA, people began to worry that the world trade system might disintegrate into regional trading blocs. The EC has a free trade agreement with the European Free Trade Association (to which several of the

European non-EC members belong) and preferential trading arrangements with Mediterranean non-member countries and with most other African countries (under the Lomé Convention). The North American countries have hinted that NAFTA could be extended to bring in South American countries. Concerned by this trend, Malaysia proposed formation of an East Asian Economic Caucus (EAEC), which could be the nucleus for an Asian trading bloc.

The threat of regionalisation is especially worrying to Australia because it could be left outside all the major blocs. Nevertheless, the danger is easy to exaggerate. The EC's 1992 program was much less dramatic than the hype suggested and was part of a continuous movement towards European integration which had a political as well as an economic dimension. Increasingly, the EC must be treated as a single unit for international trade purposes (intra-EC trade is already freer than interstate trade within some countries, e.g. China), and the real issue for non-members is the EC's external trade policies. As mentioned in other chapters of this book, these policies leave much to be desired with respect to specific goods, especially in agriculture, but there is little evidence of the EC market becoming generally less open to foreign suppliers.

The adverse consequences of NAFTA for outsiders are also easy to overstate. Although the bilateral trade flow between the USA and Canada is huge, it was already subject to few restrictions before 1989, and the vast majority of trade was subject to tariffs of less than five per cent. Extension to Mexico is proving difficult because domestic US lobbies are fearful of trade creation putting US firms out of business (or leading them to transfer operations south of the border). If a restricted NAFTA involves mainly trade diversion, then it will be less attractive to the signatories as well as encouraging membership bids from the Latin American countries whose exports are reduced. Enlargement of NAFTA will exacerbate these conflicts and probably induce it to collapse. In sum, given the absence of a political motive for integration like that driving the EC, it is difficult to envisage a comparable American Community forming an RTA covering the western hemisphere.

Within Asia there is even less cause for concern. Japan is the continent's leading economic power and, among the leading economies, it has the strongest commitment to the MFN principle. Japan is not part of any RTA and has given a cool reception to the EAEC proposal. The only active RTA in Asia is the Association of South-east Asian Nations, and in more than twenty-five years ASEAN has failed to implement any significant preferential trade policies.[4]

Australia's approach to the regionalism issue has been a precautionary one. Australia promoted the widely defined Asia Pacific Economic Cooperation (APEC), in part to forestall more restrictive groupings such as the Malaysian proposal for an East Asian Economic Caucus, which included only the ASEAN and North-east Asian countries. The idea of using APEC as a forum that also included the USA was embraced by Japan and, under Australian and Japanese leadership, it has become a vehicle for promoting open regionalism.[5] Thus, the Asia–Pacific countries will cooperate on areas of mutual interest but will not form an exclusive trading bloc.

THE URUGUAY ROUND AND THE END OF GATT

In December 1993 the Trade Negotiations Committee of GATT adopted a draft 'Final Act and Schedules' to end the Uruguay Round of multilateral trade negotiations, 2642 days after the round had been launched. This longest-running of GATT rounds was also the most complex, and the Final Act amounts to about 6500 pages of text covering a host of bilaterally negotiated agreements (which generally extend to all contracting parties under the MFN principle) as well as 'initial commitments' on liberalising trade in services in the context of a new General Agreement on Trade in Services (GATS). The negotiations also led to an agreement to establish a World Trade Organisation (WTO), which will replace the GATT Secretariat. These agreements were ratified at a ministerial meeting of the 115 contracting parties (minus Yugoslavia whose chair is vacant, and plus several observers who have membership applications on the table) in Marrakesh, Morocco, in April 1994. The WTO was scheduled to begin operations in 1995.

The Uruguay Round will lead to lower tariffs. It is, however, the first GATT round in which tariff reduction has not been central, partly because many GATT contracting parties have cut tariffs unilaterally without needing the reciprocity implicit in multilateral negotiations. During the 1980s and 1990s many Asian and Latin American countries, as well as Australia and New Zealand, substantially reduced tariffs. In the industrialised countries of North America, western Europe and Japan tariffs were already low; they will fall further, from a weighted average of 6.4 per cent to 4.0 per cent, once the Uruguay Round agreements have been implemented (envisaged to be ten years after the WTO's establishment, i.e. 2005).

The main focus of the Uruguay Round was on reducing non-tariff barriers to trade. The WTO is mandated to discipline existing and future NTBs in agriculture, in textiles and clothing and in sectors subject to voluntary export restraint agreements (notably footwear, consumer electronics and steel). The revised rules on agriculture include tariffication of all border measures (and an average reduction of agricultural tariffs by 36 per cent within six years) and reduction in domestic support and subsidised exports. The Multifibre Arrangement is to be phased out, and textiles and clothing brought back under GATT rules. Other new or stricter rules apply to trade-distorting investment incentives, laws on protection of intellectual property rights and framework agreements on services, including commitments to open markets for services.

Especially in the new areas, which are inherently harder to police than rules governing merchandise trade and for which there is little past experience, estimating the impact of the Uruguay Round package is difficult. Nevertheless, studies by the GATT Secretariat, the OECD and the World Bank suggest world income gains from freer merchandise trade of between $A400 and $A1100 billion per year. These estimates are 'static' measures, based on the models in the introduction to this book, and might be underestimates if the dynamic gains from openness suggested by the new growth theories are important (introduction, note 7). The Industry Commission has estimated that the Uruguay Round measures will boost Australian exports by more than $5 billion a year and increase GDP by around $3.7 billion, which is a substantial benefit.

There will, of course, be gainers and losers from trade liberalisation. Australian exporters will benefit substantially from the agriculture measures, for which Australian negotiators laboured long and hard. Apart from the tariffication and binding of tariffs and agreements limiting domestic supports and export subsidies for agricultural products, Australia received specific commitments from the EU and the USA not to corrupt Asian markets with subsidised beef or rice exports. In Asia, improved market access to Japanese and Korean beef markets and a permanent opening of those two countries' rice markets are especially important for Australian farmers, while dairy producers will benefit from improved access to Malaysian, Thai, Japanese, Korean, EU and US markets.[6] Increased agricultural exports will provide a significant net national gain for the country, although it is quite possible that with higher world prices Australian consumers will be paying more for their food.

The gainers outside agriculture will be the potential exporters who are hampered by other countries' trade barriers, which will include some manufacturing industries and a larger number of service activities. Specific existing industrial sectors that will benefit are coal mining, non-ferrous metals, steel, some specialised pharmaceuticals and medical and scientific equipment, processed foods and beer. Some activities will be encouraged to improve quality, e.g. agreement to protect geographical indications of wine origin will encourage Barossa Valley vintners to establish a reputation for their location (producers of Australian 'champagne' have already printed new labels). The GATS is expected to open up new opportunities for exports of telecommunications services, insurance, financial services and other business and professional services. Stronger protection of intellectual property rights will help the video and music industries (and, it is hoped, textbook writers) by outlawing pirated tapes and books, which have been a particular problem facing Australian exports to some Asian countries.

The biggest gainer of all is the Australian consumer, because Uruguay Round commitments will provide added weight to Australian governments' trade liberalisation efforts. The potential losers are workers and owners who have skills or equipment specific to the textiles, clothing and footwear industries, which will face greater competition from imports.[7] Also, producers in industries protected by anti-dumping levies or other special measures (e.g. cars) will find it harder to sustain the import barriers.

Less than half a century after the GATT was signed (as a temporary agreement to cover international trade until an International Trade Organisation was set up!), the GATT Secretariat will be replaced by the WTO. 'The new, improved and expanded version of the present GATT' (to quote an Australian Department of Foreign Affairs and Trade document publicising the Uruguay Round's outcomes for Australia) will be the forum for future international trade negotiations, allowing for continuous revisions rather than the all-or-nothing package required in GATT rounds. The WTO will prepare trade reviews, which monitor the trade practices of member countries (every two years for large trading nations, less frequently for smaller ones) and help to expose deviations from countries' international obligations. The WTO will also be responsible for a new streamlined dispute settlement mechanism, which was a major objective of smaller countries, including Australia, who suspect that large trading nations have been able to block inquiries or delay implementation of distasteful findings. Although the Uruguay Round Final Act does not have explicit rules on environmental issues, they are brought on to the future agenda by being mentioned in the preamble of the WTO text, and the need to develop a work program on trade and environment issues has been agreed.

The establishment of the WTO is viewed as heralding a new era in international trade. Many loose ends of world trade law, which could not be handled under the GATT framework and with a small secretariat, will be consolidated, and the scope of the international organisation's work will be expanded. Some observers are concerned that a larger bureaucracy might become more politicised and develop more interventionist tendencies than the old GATT, with potentially harmful consequences for a system based on liberal trade rules. That will depend primarily on the WTO member countries. The lesson of recent decades is that more and more countries appreciate the benefits of a trade system based on rights and obligations firmly rooted in international trade theory. The queue of applicants in 1994 hoping to join GATT in time to be founder members of the WTO included China and Russia, and once they have acceded no major economy will be outside the WTO. Australia has played a positive role in the recent blooming and metamorphosis of GATT and, as this section has shown, the country will reap major benefits.

CONCLUSIONS

Australia's role in international trade diplomacy has undergone a transformation during the last quarter of the twentieth century. From being a passive member of GATT for quarter of a century, Australia has become a leading actor both in the global trade negotiations and in the Asian region. During this transformation period Australian negotiators have learned that a middle power cannot impose its views, but it can successfully work to influence consensus positions so that they become more favourable.

One important lesson from the events described in this chapter is the importance of understanding the long-term consequences of trade policies. GATT has been a successful vehicle for trade liberalisation and thus contributed to global prosperity, because its articles are firmly based in trade theory: trade liberalisation is a positive-sum game with the potential to benefit all nations but, if trade barriers are used, then non-discriminatory tariffs are the least harmful. The black spots have been where countries have sought exemptions from the general principles for opportunistic reasons (e.g. on agriculture and on textiles and clothing). For Australian policymakers, the best approach to international trade diplomacy is one that emphasises commitment to the GATT principles without being seduced by apparently attractive special solutions to immediate concerns.

FURTHER READING

Curzon (1965) describes the origins and design of GATT. Oxley (1990) provides a lively guide to GATT negotiations by an Australian participant. Higgott and Cooper (1990) deal with the Cairns Group. Pomfret (1988), Anderson (1991) and Bora and Findlay (1994) cover regional trading arrangements. Garnaut's (1989) report was influential in redirecting Australian trade policies towards Asia and advocating the open regionalism approach.

205 – 08

BK Titles

EPILOGUE

Richard Pomfret

Australia

F13

Since 1973 Australia has joined in the global trend towards tariff reduction, which was started by the other high-income countries after the signing of GATT in 1947. Inspired by the huge success of the export-oriented development strategies of the East Asian newly industrialising economies, trade liberalisation spread through Latin America and Asia in the 1980s and reached Eastern Europe and the former USSR after 1989. The outlook for the next century is a positive one of an open global trading system, which will enable the world's people to benefit from the gains from specialisation and trade.

These international developments present a particular opportunity to Australia, which is sitting on the edge of the most dynamic part of the world economy and has a resource endowment that complements that of much of East and South Asia. Moreover, Australia is well placed to take advantage of new opportunities to trade in services given the emerging position of English as the international language of Asia. Rupert Murdoch's purchase of a major shareholding in the Star TV network and the entry of ABC and Channel 9 into Asian TV broadcasting in 1993 are just the most visible signs of this potential.

Behind the promise of an economic golden age lurk many dangers. On the trade policy front there are always domestic pressures from sectional interest groups to protect them from competing imports. In North America and Western Europe the post-1973 decline in economic growth was accompanied by a 'new protectionism' when grey measures like voluntary export restraint agreements flourished in contradiction to the spirit of GATT commitments. During the 1980s the belief that other countries, especially Japan, were not playing by the GATT rules led to demands for 'fair' trade policies and a level playing field, which led to aggressively unilateral actions (such as levying anti-dumping and countervailing duties) and bilateral deals to manage trade. Such measures undermine the global trading system, not least because there is disagreement over what are fair trade practices; other countries' markets always seem more difficult to crack than your own. Australia jumped on this bandwagon by imposing many anti-dumping duties during the 1980s, but far more important has been their abuse by the two big trade units. A major challenge in the coming years will be how to exercise effective international control over the use of unilateral measures by the USA and the EC. The best approach is to use international negotiations within the GATT/WTO framework to establish what are fair trade practices and to outlaw vigilante actions by individual countries.

Another perceived threat to the multilateral trading system is the emergence of regional trading blocs. Chapter 9 argued that this threat has been overstated by some alarmists, but the experience of the 1930s provides a reminder that the retreat into closed blocs is a danger if economic conditions deteriorate and nothing is done to stop a retaliatory spiral from emerging. Again, the best defence is for Australia to participate actively in the new WTO, adding its weight to those countries that oppose deviations from the unconditional most favoured nation principle.

Assuming that the omnipresent dangers of aggressive unilateralism and bilateralism are kept in check, in what direction is international trade policy likely to head? The major innovations of the Uruguay Round were to bring hitherto excluded sectors (agriculture and textiles and clothing) back into mainstream trade negotiations and to tackle new areas, such as trade in services, intellectual property rights and investment-related measures. The optimistic scenario is that all of these initiatives will be successful, and a liberal regime of international trade law will eventually cover all trade in goods and services. This will be a positive development for Australia, with its comparative advantage in many service activities and in human-capital-intensive activities where intellectual property rights are relevant.

As the direct restrictions on international trade in goods and services come down, they reveal other obstacles to improving specialisation and reaching higher social indifference curves. The analogy is often made to a swamp; only when the swamp is drained can you see the tangle of tree stumps *et cetera* that needs to be cleared away before the land can be fully utilised. The kind of stumps that need to be removed were mentioned in chapters 5 and 6; rights of establishment and peculiarities of competition policy, for example, can disadvantage foreign firms trying to operate in the Australian market or Australian firms trying to operate overseas. One predictable policy development will be the linking of trade and competition policy in international negotiations and the reaching of agreements on acceptable and unacceptable elements of national competition policies.

Other new developments will come in response to changing perceptions of society's policy goals. Here the most obvious example is the growing attention paid to environmental issues. To some extent there is still a rich country/poor country divide in this area, with countries like Brazil resisting foreign pressures to pay more attention to the environmental costs of their ambitious development goals. Nevertheless, variations in national environmental policies do distort international trade, and some environmental issues have obvious cross-border significance (e.g. when acid rain crosses into Canada or Sweden from their southern neighbours and destroys exports of maple syrup and other forest products), so that their incorporation into international agreements is inevitable. Such restrictions can do little harm to Australia, who with open spaces and clean air will benefit from increased environmental awareness (e.g. in attracting tourists).

Some of these developments have direct relevance to Australia, others do not. The overarching relevance, however, is that the world is moving towards a more open and integrated economic system, and in the last quarter of the twentieth century Australia has embraced this trend. The individual chapters of this book have argued that the domestic policy reversal was desirable to reverse Australia's poor economic growth record in the twentieth century. The challenge in the early part of the next century will be for Australia as a middle-sized economic power to use its influence to ensure that other countries, especially the economically most powerful ones, continue to play a positive role. Australian leadership in the Cairns Group has been a positive beginning, but perhaps the best approach is to inspire by example. Australia's rapid and potentially complete trade liberalisation by the end of the century, and her pioneering of wider international agreements through the CER with New Zealand, have increased her moral authority as a spokescountry for economic liberalisation. The bonus is that such international good citizenship is also in Australia's self-interest, given that past trade barriers only shot the country's citizens in their collective feet.

NA to end

ENDNOTES

INTRODUCTION

1 It will state rather than prove theorems and at several points ignore theoretically valid but (in the author's opinion) practically unimportant caveats. Many specialised texts provide a more thorough treatment. This chapter naturally follows most closely the treatment in my own text (Pomfret 1991).

2 This phrase should be repeated continually, but the significance of transport (and other transactions costs) is usually obvious and theoretically uninteresting, so in the remainder of this chapter we make the simplifying assumption that transport costs can be ignored.

3 Grubel and Lloyd (1975) favour three-digit groups of the Standard International Trade Classification (SITC) as corresponding most closely to an 'industry', but some of these groups are very heterogeneous. For Australia they found 20 per cent of trade to be intra-industry, but if industries were defined at the more disaggregated six-digit SITC level then the proportion fell to 6 per cent. How big does intra-industry trade have to be to pose a challenge to the Heckscher–Ohlin theorem? A fifth is a fair chunk, but 6 per cent is hardly worth worrying about, because no theory can explain all trade.

4 The partial equilibrium model of figure 0.5 repeats the point from the general equilibrium model of figure 0.3 that the forgone gains from trade have both a consumption and a production side component.

5 The discussion in this chapter focuses on barriers to imports, because they are by far the commonest trade restrictions. Barriers to exports can, however, be analysed with the same tools and under some conditions will have the same general equilibrium effect. This result, the Lerner symmetry

theorem, is apparent from figure 0.3, in which a tax on either imports or exports will make the domestic price line steeper than the world price line, shifting resources from the export to the import-competing activity.

6 As the economy develops, however, potential losers can insure themselves instead of relying on politically determined redistribution. In particular, as people accumulate assets, they can reduce their exposure to removal of tariff protection for the industry in which they work by buying shares in companies in other sectors.

7 The modern theory of trade and growth was pioneered by Romer (1986; 1987), Lucas (1988) and Young (1991). Grossman and Helpman (1992) is the most ambitious attempt to model the trade–growth relationship. The empirical literature on openness and growth is reviewed by Edwards (1993) and the theoretical literature by Hammond and Rodriguez-Clare (1993).

CHAPTER 1 AUSTRALIA'S CHANGING TRADE PATTERN AND GROWTH PERFORMANCE

1 Crude though these proxies are, more sophisticated indexes are unlikely to change very much the relative positions of the economies in figure 1.1.

2 The APEC grouping includes all the major trading economies of North America, East Asia and Australasia.

3 Comparisons of welfare involve more than just comparisons of GDP per capita of course. See Dowrick and Quiggan (1993) for further discussion of these data.

4 With the continuing opening of the economy the Australian import penetration and export orientation ratios have kept rising however, and in 1992–93 were 30 and 20 per cent (Industry Commission 1993:110), compared with the ratios of 24 and 13 per cent shown in table 1.11 for 1985–90.

5 The evidence from New Zealand and other countries undertaking major economic reform suggests that incomes might not increase faster until a decade or more has elapsed and only then if the reforms have been comprehensive enough as to deregulate labour as well as capital markets and services as well as goods markets (Williamson 1994).

CHAPTER 2 EVOLUTION AND POLITICAL ECONOMY OF AUSTRALIAN TRADE POLICIES

1 Similarly, the protective benefit afforded to the local substitute depends on the size of the tariffs imposed on any imported inputs used. See the discussion of effective rates of protection in the introduction.

CHAPTER 3 POLICIES AFFECTING AUSTRALIA'S PRIMARY SECTORS

1 The intra-sectoral composition of Australia's farm exports and imports would be different, however, if domestic consumer prices of protected farm products (most notably dairy products) were lowered to international prices: the lightly assisted extensive cropping and grazing activities would expand at the expense of the more-protected intensive farm activities.

2 In 1991, 86 per cent of Australians lived in urban areas (and 51 per cent in

cities of more than a million people), compared with the average for OECD countries of only 77 per cent in urban areas (and 28 per cent in cities of more than a million — see World Bank (1992:229).

3 The assisted intensive agricultural sectors such as dairying, horticulture and dried vine fruits, however, are still quite labour-intensive. They will become less so as assistance to those industries is lowered over time. For details of the myriad policy instruments used to assist these and other agricultural industries, see for example Edwards and Watson (1978), Sieper (1982) and Stanford (1994) as well as papers in Williams (1991). See also the annual reports of the Industry Commission for details of the rates of assistance on agricultural commodities.

4 A useful survey of the relevant literature on the comparative static effects of such a boom is provided by Corden (1984).

5 In the mid 1970s some of the effects of the mining boom on manufacturing output and employment were blamed on the 1973 tariff cut.

6 See, for examples, Frank Macindoe and Doug Young, 'Why the Mabo bill adds to the uncertainty of mining', *Australian Financial Review,* 23 December 1993, p 10 and Trevor Sykes, 'Hands off the panic button', *Bulletin,* 11 January 1993, pp. 16–17.

CHAPTER 4 POLICIES AFFECTING MANUFACTURING

1 Industry Commission (1992).

2 In the presence of imported intermediate goods subject to protection, the nominal rate of assistance might overstate the true amount of protection afforded to the sector. The effective rate also takes into account the impact of trade policies on a producer's input as well as output and is therefore more comprehensive, although it requires additional information on the input–output structure. It is a generalisation of the effective rate of protection concept, applied to tariffs in the introduction.

3 See Anderson and Garnaut (1987) and Capling and Garnaut (1993) for detailed studies of why the levels of protection in Australia are high relative to other developed countries.

4 See the introduction to this book and Vousden (1990, chapter 2).

5 For a tariff this is simply the per unit tariff rate multiplied by output. Note that the SE overstates the increase in producer surplus due to distorted domestic prices because it ignores the added resource cost of producing domestically units whose cost is above the world price (area b in figure 0.5).

6 See Waterson (1988) for an explanation of these concepts. Note that the CTE understates the full loss in consumer surplus, because it ignores additional units that consumers would buy if faced with the world price (area d in figure 0.5).

7 For example, in table 4.2, 1599 establishments are in the sector, yet there were only six producers of passenger motor vehicles.

8 Albon's estimates are reported in a paper, 'Export subsidisation through export facilitation: The case of passenger motor vehicles', which he

presented at the July 1992 Bureau of Industrial Economics Conference in Canberra.

9 US monthly sales peaked at 2512 in November 1991 and had dropped to 440 a year later. Some blame was ascribed to the recession-induced slump in the convertible market and to the rise of the yen (the engine and transmission are imported from Mazda in Japan), but a major problem was the poor quality of the Australian portion of the car, especially fit and finish of the body, which suggests that price alone is not crucial to export success and that the hoped-for development of an export-quality mentality did not happen (the reference here is to the personal experience of one of the authors, whose Capri runs well but the door handle has dropped off and other plastic body parts have cracked).

10 This is not to write off the Australian car industry, some of whose difficulties might be reduced by non-trade prongs of the government's economic liberalisation strategy. Labour relations have been one obstacle to productivity increases, which might have especially affected the car industry with its large workplaces and multicraft workforce. Toyota's 1991 decision to build a $240 million plant in Victoria was explicitly linked to a workplace agreement with unions, and further labour market reforms should help to bring Australian PMV factories closer into line with the international trend towards flexible workforces.

11 Another advantage to Australia is that, with global import quotas, the quota rent is less likely to accrue to foreigners than with VERs. The general point is that import quotas reduce welfare and discriminatory quotas bring additional benefits and costs, the net effect of which is ambiguous.

CHAPTER 5 SERVICE SECTOR POLICIES

Thanks to the editor for discussion on the chapter and to Geoff Peak for discussion of figure 5.7. All errors are the author's.

1 The standard categories within the sector according to the national accounting definitions are wholesale and retail trade; transport communications and storage; finance, real estate and business services; public administration and defence; community services and entertainment, personal and other services.

2 All figures in this chapter (except figure 5.6) are sourced from ABS data in the DX database. Years refer to period ending April (e.g. 1992–93 is the twelve months ending April 1993). Data for figure 5.6 were provided by Bora (1992).

3 A related difficulty is in identifying whether service sector transactions belong in GNP or GDP or both; the former measures output of goods and services by Australians, while the latter measures output of goods and services in Australia.

4 Unfortunately the data that have been used to estimate this figure include labour income flows (which are a factor service transaction and not part of the service sector — see Arndt) but they do not include capital flows.

5 Space might also be important for governments. The Singaporean air force, for example, pays to use Australian airspace for pilot training.
6 The export ratio [(X–M)/(X+M)] is confined to a range of –1 to +1 and permits a comparison of net exports over time that is independent of units of measurement. A value of zero implies that X=M. A value of –1 implies that the country only imports services. A value of +1 implies that it only exports services.
7 The Industry Commission annual reports have in recent years included an appendix on progress in microeconomic reform. The data provided there illustrate the point made here. The 1992–93 report includes examples of reforms in cargo-handling at the ports, shipping, road transport, energy supply and health services. The 1991–92 edition included details of reforms to air transport and telecommunications.
8 APEC members are Canada, the US, Mexico (since November 1993), NZ, Australia, ASEAN economies, Hong Kong, China, Taiwan, Korea and Japan.
9 The main service sector trade that so far has received attention in the complementary process in the Pacific Economic Cooperation Council (PECC) is trade in air transport services. Air transpoht is mentioned in the GATS but only with respect to ground and other complementary services, and not in relation to landing rights.

CHAPTER 6 TRADE POLICIES AND INTERNATIONAL FACTOR FLOWS

1 See the articles in the book by Maddock and McLean (1987).
2 See Goot (1988).
3 For a survey of immigration issues in the Australian context see Wooden et al. (1990).
4 See Withers (1987) and Wooden et al. (1990).
5 See the introduction and chapter 2 for discussion of the Stolper–Samuelson theorem.
6 See Brigden (1929) and Anderson (1987a).
7 The Bureau of Industry Economics (1993) reports that two surveys they have conducted confirm Brash's results.
8 See Maddock (1987) for an analysis of the long boom. Also, Pagan (1987) studies the Australian economy immediately after the boom and into the early 1980's when Australia was able to sustain a recovery.
9 All the data cited in these paragraphs are contained in United Nations (1992).
10 See Graham and Krugman (1993) for a study of the global 'surge' in DFI and Drysdale (1993) for a study of Japanese investment in Australia.
11 See Ng (1991) for an analysis of East Asian investment in Australia.
12 Defined as local sales over total sales.
13 Menzies articulated the importance of population to Australia's economic growth in a lecture at the University of Melbourne in 1962. Pope (1982) cites the Tariff Board as acknowledging that this link was still important as late as 1966.
14 His figures are 0.77 and 0.84 when the nominal rate of protection is used. All his estimates are significant at the 1 per cent level.

15 This point is reinforced by the Centre for International Economics (1988), which states that 'the problem of inefficient industries due to high tariff barriers and other protective measures is unequivocally a failure of industry policy, not immigration policy'.

CHAPTER 7 TRADE AND MACROECONOMIC POLICIES

1 Note the distinction between the real exchange rate defined as Pn/Pt and the real exchange rate defined as eP^*/P where e is the nominal exchange rate. The latter is referred to as the purchasing power parity definition and is the one most often employed in empirical macroeconomic analysis (Connolly & Devereux 1992).

2 Edwards (1986) provides an analysis of the period of transition between short- and long-run equilibrium.

3 The isocost line is defined as $TC=wL + rK$ so the slope is given by $-dw/dr= K/L$.

4 Compare the analysis of tariff-induced DFI flows in the previous chapter.

5 The fact that tariff reductions to the year 2000 were pre-announced mitigates the possibility of resource churning described in the second section of this chapter. However, the question of the credibility of government policy remains; that is, will the government continue with trade liberalisation in the face of persistent high unemployment?

6 The deterioration of the structural net export position is not to be confused with any improvement in the cyclical net export position as the economy moves into recession under the influence of high interest rates.

CHAPTER 8 TRANS-TASMAN TRADE RELATIONS

I am grateful to Peter Lloyd and to fellow contributors for comments on drafts of this chapter, and to Tony Lynch (New Zealand High Commission, Canberra), Ian Duncan (New Zealand Institute of Economic Research, Wellington) and David Round for assistance in securing research material.

CHAPTER 9 THE INTERNATIONAL POLICY ENVIRONMENT

1 Australia was a founding member of GATT and active participant in the 1955 GATT review, but refused to participate in subsequent rounds of multinational negotiations to reduce tariffs. This stance contrasted with the position of other high-income exporters of agricultural products such as Canada or Denmark, who did cut trade barriers.

2 Although Australia has received credit from her trading partners for reducing trade barriers, some practices continue to arouse criticism, especially heavy use of anti-dumping actions (Australia accounted for close to a third of all initiations reported under the GATT Anti-dumping Code in 1991–92) and some non-tariff barriers such as quarantine requirements.

3 The CER described in chapter 8 is often quoted as an RTA between like-minded nations that goes beyond narrow trade policy issues. NAFTA has also been lauded on these grounds, although some of the more complex sectoral arrangements seem designed to manage rather than liberalise trade.

4 Although ASEAN was formed in the late 1960s, attempts to establish preferential trading arrangements have faltered. In the late 1970s and early 1980s the members agreed on a program of ASEAN Industrial Projects under which each country would adopt specific projects and the other countries would agree not to duplicate the project and to source their purchases of the project's output from their ASEAN partner. In practice, commitments not to duplicate were ignored and countries purchased from the least-cost supplier rather than from the ASEAN project. Since 1992 ASEAN has been attempting to introduce a system of preferential tariffs, but national determination of goods to be included is proceeding slowly.

5 Although Australian policy makers emphasise that APEC is committed to non-discrimination, the membership of APEC suggests a regional limit to Australia's Asian orientation since APEC's western boundary cuts out much of Asia, including the fellow members of the Commonwealth in South Asia and the ten Islamic countries in the Economic Cooperation Organisation.

6 The Australian Bureau of Agriculture and Resources Economics has estimated that by 2000, when all the major accords will have been implemented, Australian farmers can expect to earn an extra $400 million on beef exports, $250 million on wheat, $170 million on dairy products, $60 million on coarse grains, $40 million on sugar, $30 million on rice and $20 million on sheepmeat due to the Uruguay Round (measured at 1993 prices).

7 Australia is not required to make any additional tariff reductions because credit is given for tariffs reduced unilaterally since the Uruguay Round began in 1986. Further cuts in protection for the TCF industries were scheduled to happen anyway, but commitments made under the Uruguay Round give future Australian governments less room for manoeuvre in the face of domestic pressure to slow the dismantling of trade barriers on TCF products.

REFERENCES

Anderson, K. (1987a), 'Tariffs and the manufacturing sector', in Maddock and McLean (1987).

Anderson, K. (1987b), 'On why agriculture declines with economic growth', *Agricultural Economics* 1(3):195–207.

Anderson, K. (1991), 'Europe 1992 and the Western Pacific economies', *Economic Journal* 101:1538–52.

Anderson, K. (1993), 'Economic growth, environmental issues and trade', in *Pacific Dynamism and the International Economic System*, eds C.F. Bergsten and M. Noland, Washington, DC: Institute for International Economics.

Anderson, K. (1994a), 'Lobbying incentives and the pattern of protection in rich and poor countries', *Economic Development and Cultural Change* 42(1) (forthcoming).

Anderson, K. (1994b), 'Multilateral negotiations, European integration and farm policy reform', *Economic Policy* 9(18):13–52.

Anderson, K. and R. Blackhurst (eds) (1992), *The Greening of World Trade Issues*, London: Harvester Wheatsheaf and Ann Arbor: University of Michigan Press.

Anderson, K. and R. Garnaut (1987), *Australian Protectionism: Extent, Causes and Effects*, Sydney: Allen & Unwin.

Anderson, K., Y. Hayami et al. (1986), *The Political Economy of Agricultural Protection: East Asia in International Perspective*, Sydney: Allen & Unwin.

Argy, F. (1990), 'Immigration and the current account deficit', in *Australia and Immigration: Able to Grow*, ed. M. Easson, Sydney: Pluto Press for the Labor Council of New South Wales.

Arndt, H.W. (1989), 'Trade in services: With special reference to ASEAN', *ASEAN Economic Bulletin* 6:1–7.

Australia (1966), *New Zealand–Australia Free Trade Agreement*, Treaty Series 1966 No. 1, Canberra: Department of External Affairs.

Australia (1983), *Australia–New Zealand Closer Economic Relations*, Treaty Series 1983 No. 2, Canberra: Department of Foreign Affairs.

Australian Bureau of Statistics (1992), *Year Book Australia*, Canberra: AGPS.

Balassa, B. (1965), 'Trade liberalisation and "revealed" comparative advantage', *Manchester School of Economic and Social Studies* 33:90–124.

Barry, F. (1992), *Market Liberalisation, Monetary Stabilisation and Foreign Debt: Did Australia Get it Wrong in the 1980s?*, Working Paper No. 21, Kensington: University of New South Wales.

Baxt, B. (1991), 'Trade prospects for Australasia–trans-Tasman competition laws', *Bulletin* (of Australia's Trade Practices Commission), No. 58 (January–February):12–16.

Bell, S. (1993), *Australian Manufacturing and the State: The Politics of Industry Policy in the Post-war Era*, Melbourne: Cambridge University Press.

Bewley, R. and G. White (1990), 'Do high interest rates improve or worsen the current account?', *Economic Papers* 9:19–32 (and see also Vol. 10, Correction, p. 92).

Bhagwati, J. (1984), 'Splintering and disembodiment of services and developing nations', *World Economy* 7:133–44.

Bhagwati, J. (1991), *The World Trading System at Risk*, London: Harvester Wheatsheaf.

Bhagwati, J. and T.N. Srinivasan (1983), *Lectures in International Trade*, Cambridge, MA: MIT Press.

Binney, J. (1987), 'Tuki's universe', in Sinclair (1987a).

Boehm, E.A. (1993), *Twentieth-century economic development in Australia*, 3rd edition, Melbourne: Longman Cheshire.

Bollard, A. and D. McCormack (1985), *Closer Economic Relations: A View from Both Sides of the Tasman,* Paper 30, Wellington: NZ Institute of Economic Research and CEDA Mongraph M79, Melbourne: Committee for Economic Development of Australia.

Bollard, A.E. and M.A. Thompson (eds) (1987), *Trans-Tasman Trade and Investment: The Effects of CER,* Research Monograph 38, Wellington: NZ Institute of Economic Research.

Bora, B. (1992), 'The impact of NAFTA on Australian foreign direct investment', Working Paper 92/10, Centre for International Economic Studies, University of Adelaide.

Bora, B. and C. Findlay (eds) (1994), *Regional Integration and the Asia-Pacific,* Melbourne: Oxford University Press.

Brecher, R. and C. Diaz-Alejandro (1977), 'Tariffs, foreign capital and immiserising growth', *Journal of International Economics* 7 (March): 317–22.

Brigden, J.B. et al. (1929), *The Australian Tariff: An Economic Inquiry,* Melbourne: Melbourne University Press.

Bureau of Industry Economics (1993), *Multinationals and Governments: Issues and Implications for Australia,* Canberra: Australian Government Publishing Service.

Bureau of Labour Market Research (1986), *Migrants in the Australian Labour Market,* Research Report No. 10, Canberra: Australian Government Publishing Service.

Bureau of Transport and Communications Economics (1991), *Costs and Benefits of a Single Australasian Aviation Market,* Canberra: Australian Government Publishing Service.

Butlin, N.G., Barnard, A. and J.J. Pincus (1982), *Government and Capitalism. Public and Private Choice in Twentieth-Century Australia,* Sydney: Allen & Unwin.

Cable, V. (1990), 'Adjusting to textile and clothing quotas: A summary of some Commonwealth countries' experiences as a pointer to the future' in *Textiles Trade and the Developing Countries,* C. Hamilton (ed.), Washington DC: World Bank, pp. 103–35.

Capling, J. and B. Galligan (1992), *Beyond the Protective State: The Political Economy of Australia's Manufacturing Industry Policy,* Melbourne: Cambridge University Press.

Carter, M. (1987), 'The service sector', in Maddock and McLean (1987).

Castle, L.V. and C. Findlay (eds) (1988), *Services Trade in the Pacific*, Sydney: Allen & Unwin.

Castles, F. (1988), *Australian Public Policy and Economic Vulnerability*, Sydney: Allen & Unwin.

Caves, R., J. Frankel and R. Jones (1990), *World Trade and Payments*, 5th ed., Boston: Little Brown.

Centre for International Economics (1988), 'The relationship between immigration and economic performance', in Report of the Committee to Advise on Australia's Immigration Policies, *Immigration: A Commitment to Australia; Consultants' Reports*, Vol. 2, Canberra: AGPS.

Chisholm, A. (1992), 'Australian agriculture: A sustainability story', *Australian Journal of Agricultural Economics* 36(1):1–30.

Cho, Y. J. (1986), 'Inefficiencies from financial liberalisation in the absence of well-functioning equity markets', *Journal of Money, Credit and Banking* 18(2):191–9.

Choski, A.M. and D. Papageorgiou (eds) (1986), *Economic Liberalisation in Developing Countries*, Oxford: Basil Blackwell.

Committee of Economic Enquiry (Vernon Committee) (1965), *Report of the Committee of Economic Enquiry*, Canberra: AGPS.

Committee to Advise on Policies for Manufacturing Industry (Jackson Committee) (1975), *Policies for Development of Manufacturing Industry*, Canberra: AGPS.

Commonwealth Bureau of Census and Statistics (various years), *Official Yearbook of the Commonwealth of Australia*, Melbourne and Canberra: Government Printer.

Commonwealth of Australia (1977), *White Paper on Manufacturing Industry*, Canberra: AGPS.

Commonwealth of Australia (Crawford Report) (1979), *Report of the Study Group on Structural Adjustment*, Canberra: AGPS.

Connolly, C. and J. Devereux (1992), 'Commercial policy, the terms of trade and real exchange rates', *Oxford Economic Papers* 44:507–12.

Corden, W.M. (1963), 'The tariff', in A. Hunter (ed.), *The Economics of Australian Industry: Studies in Environment and Structure*, Melbourne: Melbourne University Press.

Corden, W.M. (1984), 'Booming sector and Dutch disease economics: Survey and consolidation', *Oxford Economic Papers* 36(3):359–80.

Crawford, J.G. (1968), *Australian Trade Policy 1942–1966: A Documentary History*, Canberra: Australian National University Press.

Curzon, G. (1965), *Multilateral Commercial Diplomacy*, London: Michael Joseph.

Dalziel, R. (1987), ' "Misunderstandings rather than agreements": Intercolonial negotiations, 1867–1883', in Sinclair (1987a).

Davidson, F. and B. Stewardson (1979), *Economics and Australian Industry*, Melbourne: Longman Cheshire.

Donald, C.M. (1982), 'Innovation in Australian agriculture', in Williams (1982).

Doran, C.R. (1984), 'An historical perspective on mining and economic change' in *The Minerals Sector and the Australian Economy*, ed. L.H. Cook and M.G. Porter, Sydney: Allen & Unwin.

Dowrick, S. and J. Quiggin (1993), 'Australia, Japan and the OECD: GDP rankings and revealed preference', *Australian Economic Review* 101:21–34, Jan–March.

Drummond, I.M. (1974), *Imperial Economic Policy 1917–1939: Studies in Expansion and Protection*, London: Allen & Unwin.

Drysdale, P. (1993), 'Japanese direct foreign investment in Australia in comparative perspective', *Pacific Economic Papers*, No. 223, Canberra: Australia-Japan Research Centre, ANU.

Edwards, G.W. and A.S. Watson (1978), 'Agricultural Policy', in *Surveys of Australian Economics*, Vol.1, ed. F.H. Gruen, Sydney: Allen & Unwin.

Edwards, S. (1986), 'The order of liberalisation of the current and capital accounts of the balance of payments', in Choski and Papageorgiou (1986).

Edwards, S. (1989), 'On the sequencing of structural reforms', OECD Working Paper No. 70, Paris: Organisation for Economic Cooperation and Development.

Edwards, S. (1993): 'Openness, trade liberalisation and growth in developing countries', *Journal of Economic Literature* 31(1):1358–93.

Elfring, T. (1989), 'New evidence on the expansion of service employment in advanced economies', *Review of Income and Wealth* 35(4): 409–40.

Falvey, R. and N. Gemmel (1991), 'Explaining service-price differences in international comparisons', *American Economic Review* 81:1295–1309.

Findlay, C. (1990), 'Trade in services', *Asian–Pacific Economic Literature* 4:3–20.

Foreign Affairs and Trade, Department of (1993), *The APEC Region: Trade and Investment*, Canberra: DFAT.

Forsyth, P. (1993), 'Trade patterns and labour demand: International influences on wages and unemployment in Australia', paper presented to a conference on Unemployment: Causes, Costs and Solutions, Centre for Economic Policy Research, Australian National University, Canberra.

Forsyth, P. (ed.) (1992), *Microeconomic Reform in Australia*, Sydney: Allen & Unwin.

Freebairn, J.W. (1987), 'Natural resource industries', in Maddock and McLean (1987).

Gardiner, W.J. (1981), 'A colonial economy', in *The Oxford History of New Zealand*, ed. W.H. Oliver, Wellington: Oxford University Press.

Garnaut, R. (1989), *Australia and the Northeast Asian Ascendancy*, Canberra: Australian Government Publishing Service.

GATT (1993), *International Trade 1991–92*, Geneva: GATT Secretariat.

Glance, S., B. Johnston and J. Winston (1992), 'Impediments to value added minerals trade with East Asia', ABARE Research Report 92.16, Canberra: AGPS for the Australian Bureau of Agricultural and Resource Economics.

Glezer, L. (1982), *Tariff Politics: Australian Policy-making 1960–1980*, Melbourne: Melbourne University Press.

Goot, M. (1988), 'Immigrants and immigration: Evidence and argument from the polls, 1943–1987', in Report of the Committee to Advise on Australia's Immigration Policies, *Immigration: A Commitment to Australia; Consultants Reports*, Vol. 2, Canberra: AGPS.

Gould, J. (1982), *The Rake's Progress? The New Zealand Economy Since 1945*, Auckland: Hodder & Stoughton.

Graham, E. and P. Krugman (1993), 'The surge in foreign direct investment in the 1980s', in *Foreign Direct Investment Today*, ed. K. Froot, Chicago: University of Chicago Press for the National Bureau of Economic Research.

Gregory, R. (1988), 'A sad and sorry story: Australian automotive industry policy', in *International Competitiveness*, ed. H. Hazard and A.M. Spence, Boston: Ballinger Publishers.

Grilli, E.R. and M.C. Yang (1988), 'Primary commodity prices, manufactured goods prices, and the terms of trade of developing countries: What the long run shows', *World Bank Economic Review* 2(1):1–48.

Grossman, G. and E. Helpman (1992), *Innovation and Growth in the Global Economy*, Cambridge, MA: MIT Press.

Grubel, H. and P. Lloyd (1975), *Intra-industry Trade*, London: Macmillan.

Gruen, F. and M. Grattan (1993), *Managing Government: Labor's Achievements and Failures*, Melbourne: Longman Cheshire.

Hammond, P.J., and A. Rodriguez-Clare (1993): 'On endogenising long-run growth', *Scandinavian Journal of Economics* 95:391–426.

Hancock, W.K. (1930), *Australia*, London: Ernest Benn.

Harris, G.T. and F.G. Jarrett (1990), *Educating Overseas Students in Australia: Who Benefits?* Sydney: Allen & Unwin.

Hawke, G.R. (1987), 'Australian and New Zealand economic development from about 1890 to 1940', in Sinclair (1987a).

Higgott, R. and A. Cooper (1990), 'Middle power leadership and coalition building: Australia, the Cairns Group, and the Uruguay Round of trade negotiations', *International Organisation* 44:589–632.

Hill, T.P. (1977), 'On goods and services', *Review of Income and Wealth* 23:315–38.

Hirsch, S. (1989), 'Services and services intensity in international trade', *Weltwirtschaftliches Archiv* 125:45–59.

Howe, J. (1993), *Australia's Growth Potential in the 1990s*, EPAC Background Paper No. 35. Canberra: AGPS for the Office of the Economic Planning Advisory Council.

Industry Commission (1992), *Annual Report 1991–92*, Canberra: AGPS.

Industry Commission (1993), *Annual Report 1992–93*, Canberra: AGPS.

Inman, R.P. (ed.) (1985), *Managing the Service Economy: Prospects and Problems*, Cambridge and New York: Cambridge University Press.

Johnson, H. (1967), 'The possibility of income losses from increased efficiency or factor accumulation in the presence of tariffs', *Economic Journal* 77(305):151–4.

Jolly, L., T. Beck and E. Savage (1990), 'Reform of international coal trade: Implications for Australia and world trade', Discussion Paper 90.1. Canberra: Australian Bureau of Agricultural and Resource Economics.

Jupp, J. and M. Kabala (1993), *The Politics of Australian Immigration*, AGPS, Canberra.

King, S. and P.J. Lloyd (eds) (1993), *Economic Rationalism: Dead End or Way Forward*? Sydney: Allen & Unwin.

Krueger, A. O. (1986), 'Problems of liberalisation', in Choski and Papageorgiou (1986).

Krueger, A.O. (1990), 'The political economy of controls: American sugar' in *Public Policy and Economic Development: Essays in Honour of Ian Little*, M. Scott and D. Lall (eds), Oxford: Clarendon Press, pp. 170–216.

Krugman, P. and M. Obstfeld (1994), *International Economics* 2nd ed., New York: Harper Collins.

Leamer, E. (1984), *Sources of International Comparative Advantage: Theory and Evidence*, Cambridge, MA: MIT Press.

Leamer, E.E. (1987), 'Paths of development in the three-factor, n-good general equilibrium model', *Journal of Political Economy* 95(5):961–99.

Lee, C.H. and S. Naya (eds) (1988), *Trade and Investment in Services in the Asia-Pacific Region*, Boulder, CO: Westview Press.

Lloyd, P.J. (1984), 'New Zealand, CER and the Pacific', *New Zealand Economic Papers* 18:1–12.

Lloyd, P.J. (1987), 'Australia–New Zealand trade relations: NAFTA to CER', in Sinclair (1987a).

Lucas, R.E. (1988): 'On the mechanics of economic development', *Journal of Monetary Economics* 22(1):3–42.

Maddison, A. (1982), *Phases of Capitalist Development*, London: Oxford University Press.

Maddock, R. (1987), 'The long boom 1940–1970', in Maddock and McLean (1987).

Maddock, R. and I. McLean (eds) (1987), *The Australian Economy in the Long Run*, Cambridge: Cambridge University Press.

Maxcy, G. (1962), 'The motor industry', in *The Economics of Australian Industry*, ed. A. Hunter, Melbourne: Melbourne University Press.

McKinsey & Company (1993), *The Challenge of Leadership: Australia's Emerging High Value-added Manufacturing Exporters*, Melbourne: McKinsey & Company.

Michaely, M. (1992), 'The timing and sequence of trade liberalisation policy', in Choski and Papageorgiou (1992).

Mundell, R.A. (1957), 'International trade and factor mobility', *American Economic Review* 48(3):321–35.

Neary, P. and F. Ruane (1988), 'International capital mobility, shadow prices and the cost of protection', *International Economic Review* 29(4): 571–85.

New Zealand Ministry of External Relations and Trade (1992), *Closer Economic Relations: Key Documents and Speeches 1992,* Wellington.

Ng, F.H. (1991), 'Foreign direct investment from the East Asian countries and the levels of investment in Australia', BEc honours thesis, Flinders University of South Australia, Bedford Park.

Norman, N. and K. Merkle (1985), *The Economic Effects of Immigration on Australia,* Melbourne: Committee for the Economic Development of Australia.

OECD (1994) *Monitoring and Outlook of Agricultural Policies, Markets and Trade,* Paris: OECD, May.

Oliveira Martins, J. (1993), 'Market structure, international trade and relative wages', Economics Department Working Paper No. 134, OECD, Paris: OECD.

Oxley, A. (1990), *The Challenge of Free Trade,* London: Harvester Wheatsheaf.

Pagan, A. (1987), 'The end of the long boom', in Maddock and McLean (1987).

Patterson, G.D. (1968), *The Tariff in the Australian Colonies 1856–1900,* Melbourne: F.W. Cheshire.

Pinkstone, B. (1992), *Global Connections: A History of Exports and the Australian Economy,* Canberra: Australian Government Publishing Service.

Pomfret, R. (1987), 'Long-term consequences of temporary trade measures', *Challenge* 30(5):57–9.

Pomfret, R. (1988), *Unequal Trade: The Economics of Discriminatory International Trade Policies,* Oxford: Basil Blackwell.

Pomfret, R. (1991), *International Trade: An Introduction to Theory and Policy,* Oxford: Basil Blackwell.

Pomfret, R. (1992a): *International Trade Policy with Imperfect Competition,* Princeton Special Papers in Economics No. 17 (Princeton NJ), August.

Pomfret, R. (1992b) 'Imperfect competition and trade policy in small open economies' in Bureau of Industrial Economics, *1992 Conference of*

Industry Economics, Papers and Proceedings (Canberra: Australian Government Publishing Service), pp. 25–39.

Pope, D. (1982), 'Immigration and government policy: role and effects', in *The Economics of Australian Immigration*, ed. D. Douglas, Sydney: Sydney University Press.

Rattigan, G.A. (1986), *Industry Assistance: The Inside Story*, Melbourne: Melbourne University Press.

Reitsma, A.J. (1960), *Trade Protection in Australia*, Brisbane: University of Queensland Press.

Report of the Committee to Advise on Australia's Immigration Policies, *Immigration: A Commitment to Australia; Consultants Reports*, Vol. 2, Canberra: AGPS.

Reserve Bank of Australia, *Australian Economic Statistics 1949–50 to 1989–90*, Reserve Bank of Australia Occasional Paper No. 8, n.d.

Ricardo, D. (1817), *On the Principles of Political Economy and Taxation*, London: John Murray.

Rogers, C. and P. Neal (1993), *Macroeconomics and the Australian Economy*, Sydney: Prentice-Hall.

Romer, P.M. (1986): 'Increasing returns and long run growth', *Journal of Political Economy* 94(5):1002–38.

Romer, P.M. (1987), 'Growth based on increasing returns due to specialisation', *American Economic Review (Papers and Proceedings)* 77:56–62.

Ross, A. (1990), 'Australian overseas trade and national development policy 1932–1939: A story of colonial larrikins or Australian statesmen?' *Australian Journal of Politics and History* 36(2):184–204.

Sampson, G. and R. Snape (1985), 'Identifying the issues in trade in services', *World Economy* 8:171–82.

Shann, E. (1930), *An Economic History of Australia*, Cambridge: Cambridge University Press. (Reprinted Melbourne: Georgian House, 1963.)

Sieper, E. (1982), *Rationalising Rustic Regulation*, Sydney: Centre for Independent Studies.

Simkin, C.G.F. (1951), *The Instability of a Dependent Economy: Economic Fluctuations in New Zealand 1840–1914*, London: Oxford University Press.

Sinclair, K. (1987a), *Tasman Relations: New Zealand and Australia 1788–1988*, Auckland: Auckland University Press.

Sinclair, K. (1987b), 'Why New Zealanders are not Australians: New

Zealand and the Australian federal movement, 1881–1901', in Sinclair (1987a).

Sinclair, K. (1987c), 'The great Anzac plant war: Australia–New Zealand trade relations 1919–39', in Sinclair (1987a).

Siriwardana, A.M. (1991), 'The impact of tariff protection in the colony of Victoria in the late nineteenth century: A general equilibrium analysis', *Australian Economic History Review* 31(2):45–65.

Smith, B. (1983), 'Resources and Australian economic development', in *Surveys of Australian Economics,* Vol. 3 (ed. F.H. Gruen), Sydney: Allen & Unwin.

Smith, J.P. (forthcoming), *The History of Economic Policy in Australia,* Melbourne: Longman Cheshire.

Spenceley, G. (1990), *A Bad Smash: Australia in the Depression of the 1890s,* Melbourne: McPhee Gribble.

Stanford, L.J. (1994), 'The effects of regulation and assistance in the Queensland raw sugar industry', Policy Discussion Paper 94/01, Centre for International Economic Studies, University of Adelaide.

Steenblik, R.P. and K.J. Wigley (1990), 'Coal policies and trade barriers', *Energy Policy* 18(4):351–69.

Stern, R.M. and B.M. Hoekman, (1988), 'The service sector in economic structure and in international transactions', in Castle and Findlay (1988).

Stolper, W.F. and P.A. Samuelson (1941), 'Protection and real wages', *Review of Economic Studies* 9(1):58–73.

Stubbs, P. (1972), *The Australian Motor Industry: A Study in Protection and Growth,* Melbourne: Longman Cheshire.

Swan, P. and J. Zeitsch (1992), 'The emerging Australian manufacturing export response to microeconomic reform', *Australian Journal of Management* 17:21–66.

Thomas, M. (1988), 'Manufacturing and economic recovery in Australia, 1932–1937', in *Recovery from the Depression: Australia and the World Economy,* R.G. Gregory and N.G. Butlin, eds, Melbourne: Cambridge University Press.

Toh, M.H. and L. Low (1989), 'Singapore's service sector development in ASEAN context', *ASEAN Economic Bulletin* 6:8–30.

Tsokhas, K. (1992), ' "A pound of flesh": war debts and Anglo-Australian relations, 1919–1932', *Australian Journal of Politics and History* 38(1):12–26.

Tyers, R. and K. Anderson (1992), *Disarray in World Food Markets: A Quantitative Assessment*, Cambridge: Cambridge University Press.

United Nations (1992), *World Investment Report: Transnational Corporations as Engines of Growth*, New York: United Nations Centre on Transnational Corporations.

Valentine, T. (1987), 'The depression of the 1930s', in Maddock and McLean (1987).

Vamplew, W. (ed.) (1987), *Australian Historical Statistics*, Sydney: Fairfax, Syme & Weldon Associates.

Villanueva, D. and A. Mirakhor (1990), 'Strategies for financial reforms', *IMF Staff Papers* 3:509–36.

Vousden, N. (1987), 'Content protection and tariffs under monopoly and competition', *Journal of International Economics* 23:263–82.

Vousden, N. (1990), *The Economics of Trade Protection*, Cambridge: Cambridge University Press.

Waterson, M. (1988), 'An outsider's view of the Button Plan for the car industry', *Economic Papers* 17:19–30.

Williams, D.B. (ed.) (1991), *Agriculture in the Australian Economy* (3rd ed.), Melbourne: Oxford University Press (2nd ed., 1987).

Williamson, J. (ed.) (1994), *The Political Economy of Policy Reform*, Washington, DC: Institute for International Economics.

Withers, G. (1987), 'Labour', in Maddock and McLean (1987).

Wooden, M. (1994), 'The economic impact of immigration', in Wooden et al. (1994).

Wooden, M., R. Holton, G. Hugo, J. Sloan (eds) (1994), *Australian Immigration: A Survey of the Issues*, 2nd ed., Canberra: Bureau of Immigration Research.

World Bank (1993a), *World Tables* 1992–93, Baltimore: Johns Hopkins University Press.

World Bank (1993b), *World Development Report* 1993, New York: Oxford University Press.

Young, A. (1991), 'Learning by doing and the dynamic effects of international trade', *Quarterly Journal of Economics* 106(2):369–406.

Index